Inelastic X-Ray Scattering and X-Ray Powder Diffraction Applications

Edited by Alessandro Cunsolo, Margareth K. K. D. Franco and Fabiano Yokaichiya

Published in London, United Kingdom

IntechOpen

Supporting open minds since 2005

Inelastic X-Ray Scattering and X-Ray Powder Diffraction Applications
http://dx.doi.org/10.5772/intechopen.77434
Edited by Alessandro Cunsolo, Margareth K. K. D. Franco and Fabiano Yokaichiya

Contributors
Toshiji Kanaya, Makina Saito, Ryo Mashita, Takeshi Egami, Alessio De Francesco, Alessandro Cunsolo, Luisa Scaccia, Adriana Da Cunha Rocha, Gabriela Ribeiro Pereira, Eva Maria Perez Soriano, Isabel Montealegre-Meléndez, Cristina M. Arevalo Mora

First published in London, United Kingdom, 2020 by IntechOpen
IntechOpen is the global imprint of INTECHOPEN LIMITED, registered in England and Wales, registration number: 11086078, 7th floor, 10 Lower Thames Street, London, EC3R 6AF, United Kingdom
Printed in Croatia

British Library Cataloguing-in-Publication Data
A catalogue record for this book is available from the British Library

Additional hard and PDF copies can be obtained from orders@intechopen.com

Inelastic X-Ray Scattering and X-Ray Powder Diffraction Applications
Edited by Alessandro Cunsolo, Margareth K. K. D. Franco and Fabiano Yokaichiya
p. cm.
Print ISBN 978-1-78985-051-2
Online ISBN 978-1-78985-052-9
eBook (PDF) ISBN 978-1-83968-494-4

For EU product safety concerns:
IN TECH d.o.o., Prolaz Marije Krucifikse Kozulić 3, 51000 Rijeka, Croatia, info@intechopen.com or visit our website at intechopen.com.

Meet the editors

Alessandro Cunsolo is a Physicist at National Synchrotron Light-II (NSLS-II) at Brookhaven National Laboratory, Upton, New York. He has been working in university laboratories for four years and in national laboratories and research facilities for 21 years, both in Europe (*Institut Laue Langevin* and *European Synchrotron Radiation Facility*, Grenoble, France) and the United States (*Advanced Photon Source at Argonne National Laboratory*, Lemont, IL, and *NSLS-II*). He's an acknowledged expert in the physics of liquids and amorphous systems and in the development of inelastic x-ray and neutron scattering spectrometers in synchrotron and reactor-based research facilities. In his research activity, he has accrued expertise in different experimental methods of dynamic characterization of samples, also including neutron and x-ray, UV and visible light probes. His research achievements are documented by about 100 publications in scientific journals.

Dr. Franco is a researcher at the Institute of Energy and Nuclear Research, Department of Radioactive Rejects in Brazil. She graduated from EEL-USP São Paulo University in 1982 and received her MSc in Chemical Engineering from the University of Campinas, UNICAMP-Brazil, in 1987. She received her PhD in Electrical Engineering from University of Campinas with the thesis entitled *Deposition and Characterization of semiconductor diamond films*, in 1996. Dr. Franco has published papers in local and international peer-reviewed journals in the fields of cementitious materials, biochemistry, and pharmacology. She has worked for the Brazilian Synchrotron Light Laboratory, X-ray Powder Diffraction Beamline, dedicated to meeting the needs of researchers from several countries, providing technical and scientific support. She sustained collaborations with researchers from Germany and Australia to study green cement, composites with cement and irradiated cement. Dr. Franco's research with drug delivery systems focuses on characterization through small angle x-ray scattering and small angle neutron scattering.

Dr. Yokaichiya was a researcher and an instrument scientist for the 2-axis diffractometer (E4) and for the flat cone diffractometer (E2) at Helmholtz-Zentrum Berlin für Materialen und Energie (HZB), Germany. He graduated as an Electronic Engineer from the Federal University of Parana (UFPR- Brazil) in 1995, and received his PhD degree in Physics at the University of Campinas (UNICAMP- Brazil) in 2003. With his PhD research, he studied strong correlated systems using synchrotron radiation facilities. From 2005 to 2010 he worked as a post-doctorate in several laboratories: Laboratoire Louis Néel in Grenoble (France, 2005), Brookhaven National Laboratoy (USA, 2006), and Helmholtz-Zentrum Berlin (Germany, 2007-2010). From 2010 to 2013, Dr. Yokaichiya worked at LNLS as a beamline scientist responsible for the x-ray powder diffraction beamline (XPD). His main projects focus on magnetic materials, drug delivery systems, and cement/concrete systems.

Contents

Preface

The characterization of microscopic properties of materials over distances comparable with, or smaller than, the nearest atoms' separations and time-lapses roughly matching "in-shell" rattling periods has become of critical relevance in our society and has been the focus of intensive research, having both fundamental and applicative aspects. From the experimental side, a relevant portion of these studies is routinely carried out using large-scale X-ray of neutron facilities research and relies on the use of sources, as well as more conventional X-ray instrumentation.

Rather than providing a comprehensive overview of scientific opportunities offered by this field of research, this book aims to give the reader a taste of a few novel applications of two complementary scattering methods: high-resolution Inelastic X-ray Scattering (IXS) and Powder Diffraction (XPD), respectively characterizing dynamic (time-dependent) and static (structural) properties of materials at mesoscopic scales.

In a scattering measurement, a beam of particles-waves impinges on a sample at thermal equilibrium exchanging with it both **energy** and **momentum**, and being, as a result, scattered in all directions. If the perturbation induced on the target sample is weak, the dependence of the scattered beam intensity on the energy and momentum exchanged is uniquely informative of equilibrium properties of the sample. When the beam's particles involved are X-rays, such a scattering intensity ultimately conveys information on the positions and the movements of the target sample's molecules.

Conceptually, a scattering instrument resembles a microscope pointed on either the structure or the dynamics of the target sample. Indeed, it can be zoomed in or out to focus on various distances and time-lapses by a suitable variation of momentum and energy transfers, respectively. Since the latter variables both increase upon decreasing the incident beam wavelength, the use of short wavelength radiation, as X-rays, can shed light phenomena occurring over extremely short distances and timescales.

It is customary to distinguish between diffraction and inelastic scattering methods: the first elucidates static properties of materials through the scanning of the exchanged momentum, while the second offers the additional option of varying the frequency, as required to investigate dynamic phenomena.

As mentioned, this book mainly deals with two specific examples of inelastic scattering and a diffraction technique (IXS and XPD) and its chapters are organized as follows.

The first, introductory chapter outlines relevant analytical steps toward a formal expression of the IXS cross-section, eventually demonstrating its direct link with spontaneous density fluctuations in fluids at equilibrium. It also provides an estimate of the count rate achievable in typical IXS measurements, succinctly comparing the outcome of an IXS measurement with a similar determination achieved by the complementary terahertz technique, Inelastic Neutron Scattering (INS).

Even from this introductory chapter, it readily appears that the IXS signal from disordered materials has a nearly structureless shape, whose interpretation is often hampered by a limited energy resolution and count statistics accuracy. Not uncommonly, these inherent difficulties are overlooked when analyzing the measured lineshape, while assuming overly invasive and inherently biased hypotheses on the analytical form of such a profile. Furthermore, these models are sometimes arbitrary or contain an unreasonably large number of free parameters. This course of events makes especially critical the need for a probabilistically grounded modeling of the lineshape. A substantial improvement can be achieved by implementing Bayesian inference methods, as discussed in Chapter 2. This chapter shows how Bayesian inference principles can be used to perform hypothesis tests involving competitive lineshape models. This approach inherently embodies, as a selection criterion, the "Occam razor" principle, which states that, among alternative explanations of some evidence, the one containing less adjustable parameters is always preferable.

Overall, the chapter illustrates the benefits of this approach for the interpretation of both frequency and time-resolved scattering results. This inherent versatility is of special value for IXS, which can also be implemented as time-domain spectroscopy, as discussed in Chapter 3. This chapter deals with the representation of IXS results in the direct (space-time) domain, rather than in the more conventional reciprocal (frequency-wave vector) one. Specifically, it shows how this representation makes the interpretation of results more straightforward. Chapter 4 illustrates the potentialities of ultra-high-resolution quasi-elastic Mössbauer gamma-ray spectroscopy with energy resolution in the neV-window. This unique performance enables the study of the microscopic dynamics over timescales included between nanoseconds and microseconds. Results can be obtained either in the time or the energy domain using either a time-domain interferometer or a nuclear Bragg monochromator, respectively.

As mentioned, the second technique dealt with in this book is the X-Ray Powder Diffraction (XPD), presented in Chapters 5 and 6. These chapters focus on the characterization of manufacturing materials using conventional X-ray laboratory instruments. Chapter 5 explores the XPD from steels to identify and quantify the phases, using the Rietveld method, a method potentially applicable in industrial environments. Chapter 6 illustrates in situ XPD results, aiming at studying the formations of secondary phases in titanium composite materials under the influence of the fabrication parameters.

In summary, we believe that this book is a useful reference for those who want to use these techniques to improve the current knowledge of microscopic properties of materials.

Alessandro Cunsolo
Department of Physics,
University of Wisconsin–Madison,
United States

Margareth K. K. D. Franco
Instituto de Pesquisas Energéticas e Nucleares-IPEN,
Cidade Universitária,
São Paulo (SP), Brazil

Fabiano Yokaichiya
Universidade Federal do Paraná,
Curitiba (PR), Brazil

Section 1

Inelastic X-Ray Scattering

Chapter 1

High-Resolution Inelastic X-Ray Scattering: A Probe of Microscopic Density Fluctuations in Simple Fluids

Alessandro Cunsolo

Abstract

The explicit form of the inelastic X-ray scattering, IXS, cross-section is derived within a time-dependent perturbative treatment of the scattering process. In this derivation, the double differential cross-section is obtained from the Fermi Golden Rule within a plane wave expansion of the vector potential. Furthermore, it is assumed throughout that the Thompson term of the perturbative Hamiltonian yields the overwhelming contribution to the scattering. The achievement of an explicit form for the double differential scattering cross-section rests on the validity of the adiabatic or Born-Oppenheimer approximation. As a result, it is here shown that that the IXS double differential cross-section is proportional to the spectrum of density fluctuations of the sample, which is thus the sample variable directly accessed by IXS measurements. Although the whole treatment is valid for monatomic systems only, under suitable approximations, it can be extended to molecular systems.

Keywords: inelastic X-ray scattering, theory of the scattering, theory of the line-shape, double differential scattering cross-section

1. Introduction

Inelastic scattering measurements are among the most powerful tools to investigate the collective terahertz dynamics of disordered systems [1, 2]. Although this subject has been the focus of intense scrutiny in the past few decades, it still presents many challenging aspects. In a spectroscopic measurement, the dynamic response of the target system is stimulated via the exchange of an energy $\hbar\omega$ and momentum $\hbar\mathbf{Q}$ where $\hbar$ is the reduced Planck's constant. A suitable choice of the exchanged wavevector amplitude $Q = |\mathbf{Q}|$ and ω enables to tune the probe to dynamic events occurring over different scales. For infinitesimal Q and ω values, the measurement probes slowly decaying, hydrodynamic, density fluctuation modes either propagating or diffusing throughout the system, which resembles a continuous and homogeneous medium [3]. Upon increasing Q's and ω's, probed dynamic events become gradually faster and involve fewer atoms until the extreme, single-particle limit is reached. In this limit, the probe couples with the free recoil of the single atom after the collision with the photon and before any interaction with the first neighboring atomic cage [4].

Although the spectral profile is exactly known analytically in both hydrodynamic and single-particle limits, its evolution at the crossover in between them still eludes a firm understanding. Particularly insightful appears the study of the lineshape in the mesoscopic range, which corresponds to $2\pi/Q$ and $2\pi/\omega$ values roughly matching nearest neighbor separations and 'in cage' rattling periods of atoms, respectively.

This range is the natural domain of high-resolution inelastic scattering, IXS [5], a spectroscopic method, which, since its development towards the end of the past millennium, has substantially improved the current understanding of the terahertz dynamics of condensed matter systems. This success partly owes to both inherent and practical advantages that this technique offers compared to the complementary terahertz spectroscopy, inelastic neutron scattering, INS. Intrinsic benefits include the virtual absence of kinematic limitations, the straightforward implementation of constant-Q energy scans, a mostly coherent cross-section and an often negligible multiple scattering contribution. More practical strengths are instead the substantially higher photon fluxes impinging on the sample and the smaller transversal size of the beam. However, these undoubted advantages can only be obtained at the cost of substantial count rate penalties. Indeed, the investigation of the collective dynamics in disordered systems imposes the access to energy transfer $E = \hbar\omega$ as low as a few meV. For IXS spectrometers, typically operated at 2.1 10^4 eV, resolving those energies imposes a resolving power of $\Delta E/E \leq 10^{-7}$. The achievement of such a challenging performance has held back for long the development of high-resolution IXS, which was only made possible by the advent of high-brilliance third-generation synchrotron sources and by parallel advances in the X-ray optics [6, 7].

As an introduction to the field, this chapter is devoted to a derivation of the cross-section of IXS measurements, thus elucidating its direct connection with the Fourier transform of the atomic density fluctuations autocorrelation function. A similar treatment, which can also be found in Refs. [5, 8], is strictly valid for monatomic systems only, even though it can be easily generalized to the case of molecular systems.

2. Generalities on an inelastic scattering measurement

In a typical IXS measurement, a beam of particles–waves, as, for example, neutrons, X-rays or electrons, having well-defined energy, wavevector and polarization impinges on a sample and, after the impact, it is scattered all over the solid angle. A detector placed at a distance r from the sample is used to count the particles deviated by an angle 2θ within the small solid angle $\Delta\Omega$ and intercepting its sensitive area $A = r^2\Delta\Omega$. Along the whole flight from the source to the detector, photons pass through optical elements filtering their energy both upstream and downstream of the sample, respectively referred to as monochromators and analysers. Other devices, such as collimators, mirrors, compound reflective lenses and so forth, are commonly used to shape the particle beam as required by experimental needs, and, specifically, they define its angular divergence and, whenever needed, its polarization.

At a long distance from the centre of the scattering, the electromagnetic wave generated by the scattering event is the sum of a plane and a spherical wave [9], that is, waves having respectively a planar and a spherical wavefront. In other terms, the ultimate effect of the scattering source is to remove a part of the photons from their initial 'plane wave' state and reradiate them into a spherical wave, which is consistent with well-known Huygens-Fresnel principle [10]. However, it is always safe to assume that the detector—which intercepts photons deviated by an angle 2θ—has sensitive area $A = r^2\Delta\Omega$ small enough to safely approximate the spherical wave impinging on it as a plane wave. As a consequence, the scattering event probed in a real experiment can be portrayed as a transition of the photon states between two different plane waves, as schematically shown in **Figure 2**. These are characterized by well-defined wavevector $\boldsymbol{k}_{i,f}$ energy, $\hbar\omega_{i,f}$ and polarization $\hat{\varepsilon}_{i,f}$, with the indices '*i*' and '*f*' labelling the initial and final values, that is, the values before and after the scattering, respectively. Here all vector variables are indicated in bold.

As it appears from **Figure 1**, if one considers the plane defined by the two vectors $\boldsymbol{k}_i$ and $\boldsymbol{k}_f$, only one angular coordinate, the scattering angle 2θ, is sufficient to describe the scattering problem.

To derive an expression of the intensity detected in high-resolution *IXS* measurements, it is useful to recognize that these measurements are typically executed in transmission geometry, that is, by detecting the scattering signal downstream of the sample.[1]

For the sake of simplicity, a few more assumptions are here considered: (1) the sample has a straightforward shape: a slab of thickness t_s; (2) such a slab is crossed by the incident beam orthogonally to its front area; (3) the beam cross-section Σ_B is constant throughout the sample thickness, which implies that we are discarding the focusing of the incident beam; and (4) finally, for most *IXS* measurements, one can further assume that the beam only illuminates a limited portion of the whole cross-sectional area of the sample. However, the detector has a sensitive area sufficiently

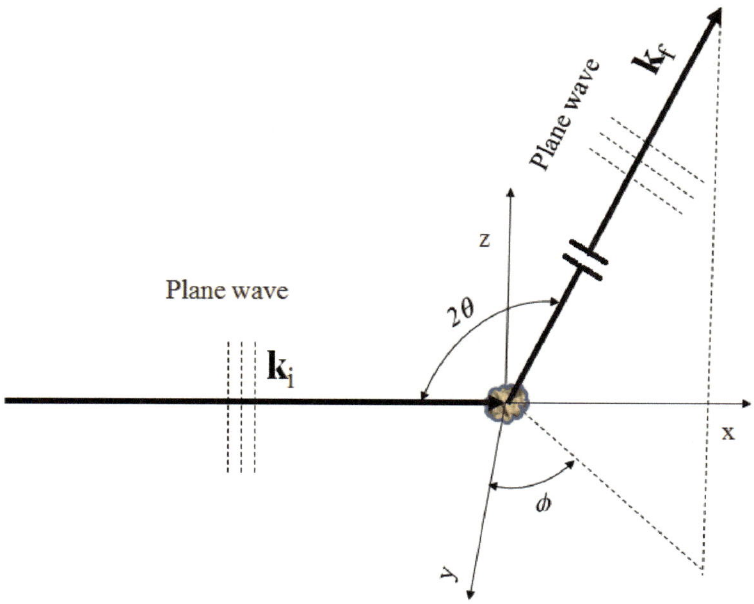

Figure 1.
A schematic rendering of the scattering process and the plane wave approximation (see text).

[1] For simple sample shapes, the treatment can be easily extended to the case of finite scattering angle by using simple trigonometry.

small, and a distance from the sample sufficiently large, that the scattered radiation impinging on it is schematizable as a plane wave, having wavevector $\boldsymbol{k}_f$ and wavefront perpendicular to it.

Under these assumptions, we can write a general expression to estimate the number of photons per unit time impinging on the detector, which is given by:

$$I \propto \Phi n_s \Sigma_B t_s \left(\frac{\partial^2 \sigma}{\partial \Omega \partial E_f} \right) \Delta \Omega dE_f, \tag{1}$$

where Φ is the photon flux on the sample, defined as the number of photons impinging on the sample per unit time and unit area, while n_s is the number of scattering units per unit volume, which is here assumed constant throughout the X-ray-illuminated sample.

The above formula introduces the double differential scattering cross-section:

$$\frac{d^2\sigma}{d\Omega\, \mathrm{d}E_f} = \frac{\begin{array}{c}\text{Rate of photons scattered into } d\Omega \text{ with} \\ \text{final energy between } E_f \text{ and } E_f + dE_f\end{array}}{\Phi d\Omega\, \mathrm{d}E_f}, \tag{2}$$

which is the only parameter of Eq. (1) conveying non-trivial information on the sample properties.

It can be recognized that the beam intensity across the sample thickness is not constant, as a part of it gets absorbed by the sample itself. This intensity reduction can be easily evaluated by expressing the attenuation caused by an elemental sample slice of thickness dx and located at a distance x. This intensity loss reads as:

$$dI = I(x + dx) - I(x) = -I\mu dx, \tag{3}$$

where μ is the absorption coefficient at the energy of the incident beam. The integration of both members of the above equation leads to the conclusion that the intensity transmitted through the sample experiences an exponential decay. Assuming a forward scattering geometry, the attenuation factor can be simply obtained as $exp\,(-\mu t_s)$ and inserted in Eq. (1), thus obtaining:

$$I = \Phi n_s \Sigma_B t_s\, exp\,(-\mu t_s) \left(\frac{\partial^2 \sigma}{\partial \Omega \partial E_f} \right) \Delta \Omega dE_f. \tag{4}$$

The above formula lends itself to a direct estimate of the ideal sample thickness, which is identified by the $\partial I / \partial t_s = 0$ condition, which yields $t_s = 1/\mu$. In summary, the optimal sample thickness should match the absorption length of the sample at the energy of the incident beam. For typical incident beam energies of most current *IXS* spectrometers, and for sample atomic species having electron number $Z > 4$, the extinction of the incident intensity is primarily caused by the photo-electric absorption process, which dominates over the Thomson scattering. The photoabsorption length typically decreases upon increasing Z, and this implies that *IXS* measurements on low Z materials require the use of relatively large samples, with a thickness in the *cm* range. However, this requirement becomes prohibitive for samples available in a small amount or that must be embedded in small volumes, as is typically the case of high-pressure experiments in Diamond Anvil Cells, *DACs*.

3. The interaction between impinging electromagnetic field and target electrons

Given this preliminary discussion, the focus is now on the analytical derivation of the *IXS* double differential cross-section $d^2\sigma/d\Omega dE_f$. An explicit analytical form requires, in the first place, a suitable expression for the Hamiltonian describing the interaction between the impinging photon beam and the electrons of the target sample. If one discards the relativistic nature of electron movements and neglects the usually weak contribution from the electron spin, such a Hamiltonian has the following form [5]:

$$H = \frac{1}{2M_e}\sum_i \left[\boldsymbol{p}_i - \frac{e}{c}\boldsymbol{A}(\boldsymbol{r}_i)\right]^2 + \sum_i V(\boldsymbol{r}_i) + V_{\text{int}}^{e-e}, \tag{5}$$

where $\boldsymbol{r}_i$ and $\boldsymbol{p}_i$ are the position and the momentum of the ith electron, respectively, V_{int}^{e-e} is the electron–electron interaction potential averaged over the electron clouds of target atoms, while $V(\boldsymbol{r}_i)$ is the potential acting on the ith electron. The above Hamiltonian can be cast in the following perturbative form:

$$H = H_{\text{el}} + H_{\text{int}}^{(1)} + H_{\text{int}}^{(2)}, \tag{6}$$

where the unperturbed Hamiltonian, associated with the multielectron system in the absence of the electromagnetic field, reads as:

$$H_{\text{el}} = \sum_i \left[\frac{p_i^2}{2M_e} + V(\boldsymbol{r}_i)\right] + V_{\text{int}}^{e-e}, \tag{7}$$

plus the other two terms accounting for the perturbation induced by the impinging electromagnetic field, that is, respectively:

$$H_{\text{int}}^{(1)} = \frac{-e}{2M_e c}\sum_i \{\boldsymbol{A}(\boldsymbol{r}_i), \boldsymbol{p}_i\} \tag{8}$$

and the so-called Thomson scattering term:

$$H_{\text{int}}^{(2)} = \frac{1}{2} r_0 \sum_i \boldsymbol{A}(\boldsymbol{r}_i) \cdot \boldsymbol{A}(\boldsymbol{r}_i). \tag{9}$$

Here the symbol $\{,\}$ denotes the anticommutator operator, while $r_0 = e^2/(M_e c^2)$ is the classical electron radius expressed in cgs units. To its leading order, the perturbation $H_{\text{int}}^{(1)}$ in Eq. (8) describes one-photon interactions with the sample as absorption and emission, while two-photon processes, such as the scattering event, come into play to the second-order only. Conversely, the Thomson term (Eq. (9)), being quadratic in the vector potential, accounts to the first order for two photons interactions such as the scattering event. Away from an energy resonance, the latter term largely exceeds the second-order expansion of Eq. (8), thus providing an overwhelming contribution to the scattering process, which will be hereafter assumed to be entirely described by the Thomson term.

As mentioned, in a typical scattering measurement, the X-ray photons undergo a transition between two different plane wave states. Therefore, one could, in principle, use the Fermi Golden Rule [11] to count all scattered photons emanating from a single incident plane wave and having wavevector pointing to a 2θ direction to within a solid angle $\Delta\Omega$, thus deriving the double differential cross-section explicitly.

This strategy would require, in principle, a proper normalization of the photon wave functions, but, unfortunately, plane waves have normalization integral diverging for long distances. This difficulty is usually circumvented by confining the description of the scattering problem to a cubic box of size L and eventually considering the limit for large L. Within this L-sized cubic box, the vector potential becomes a linear combination of normalized plane waves which explicitly reads as [5]:

$$\boldsymbol{A}(\boldsymbol{r}) = \sum_{k,\alpha}\sqrt{\left(\frac{2\pi\hbar}{\omega_k L^3}\right)}c\hat{\varepsilon}_\alpha\left[a_{k,\alpha}\exp\left(i\boldsymbol{k}\cdot\boldsymbol{r}\right) + a^\dagger_{k,\alpha}\exp\left(-i\boldsymbol{k}\cdot\boldsymbol{r}\right)\right]. \tag{10}$$

Here the indexes 'k' and 'α' label, respectively, the wavevector and the polarization states of the wave; $a_{k,\alpha}$ and its Hermitian conjugate $a^\dagger_{k,\alpha}$ are the annihilation and creation operators, respectively; c is the speed of light in vacuum and ω_k is its angular frequency. Notice that the plus and minus signs in the phases of the exponential terms of Eq. (10) respectively define the upstream and downstream propagation of the photon plane wave.

Coming back to the double differential scattering cross-section, one can express it as:

$$\frac{d^2\sigma}{d\Omega dE_f} = \frac{dP_{i\to f}}{dt}\frac{1}{\Phi}\frac{d^2n}{d\Omega dE_f}, \tag{11}$$

where $dP_{i\to f}/dt$ is the probability rate per sample and probe units that a photon experiences a transition between the initial and the final photon states, while the term $d^2n/d\Omega dE_f$ represents the density of final photon states. The probability rate in Eq. (11) should be more appropriately written as a sum over all elementary excitations in the sample possibly coupling with the scattering event. Hence,

$$\frac{dP_{i\to f}}{dt} = \sum_{I,F}\frac{dP_{I,i\to Ff}}{dt}, \tag{12}$$

with $P_{I,i\to Ff}$ denoting the probability of a transition $|I,i\rangle \to |F,f\rangle$ between the combined states of the photon and the sample, labeled by lower case and capital fonts, respectively.

Eq. (12) is particularly useful as the term under summation can be derived explicitly using the Fermi Golden Rule, according to which:

$$\frac{dP_{I,i\to Ff}}{dt} = \frac{2\pi}{\hbar}\left(\frac{d^2n}{d\Omega dE_f}\right)|\langle F,f|H_{int}|I,i\rangle|^2. \tag{13}$$

The last factor in the right-hand side of the above equation contains the perturbative part of the Hamiltonian computed between initial and final combined photon and sample states. As mentioned, we will assume that this term entirely coincides with the Thomson term in Eq. (9).

At this stage, the derivation of the double differential cross-section requires one to tackle the density of final states $d^2n/d\Omega dE_f$ analytically, as discussed in the next paragraph.

4. Counting the photon states

It is worth noticing that the expedient of circumscribing the scattering within a L-sized cubic box, besides enabling a proper normalization of the plane waves, makes more straightforward the counting of the final state photon modes [11]. The number of plane waves with energy included between E_f and $E_f + dE_f$ and pointing to a direction 2θ within a solid angle $\Delta\Omega$ is given by:

$$\left(\frac{d^2n}{d\Omega dE_f}\right)dE_f\Delta\Omega. \tag{14}$$

In the reciprocal space, the bandwidth dE_f corresponds to the volume $dV(k_F)$ of the spherical shell of infinitesimal thickness represented in **Figure 2**, for which one can write $dV(k_f) = d\Omega k_f^2 dk_f$. The wavevectors' components in the box L representing the boundary of our scattering problem are:

$$k_x = (2\pi/L)n_x \;\; k_y = (2\pi/L)n_y \;\;\; k_z = (2\pi/L)n_z, \tag{15}$$

where n_x, n_y and n_z are generic integers.

The set of wavevectors defined in Eq. (15) identifies a lattice in k-space, whose simplest self-replicating unit cell has a volume $V_{\min} = (2\pi/L)^3$. If this volume is small enough—or, equivalently, if L is large enough—the number of lattice points within the elemental volume is given by the ratio $dV(k_f)/V_{\min}$ (**Figure 2**).

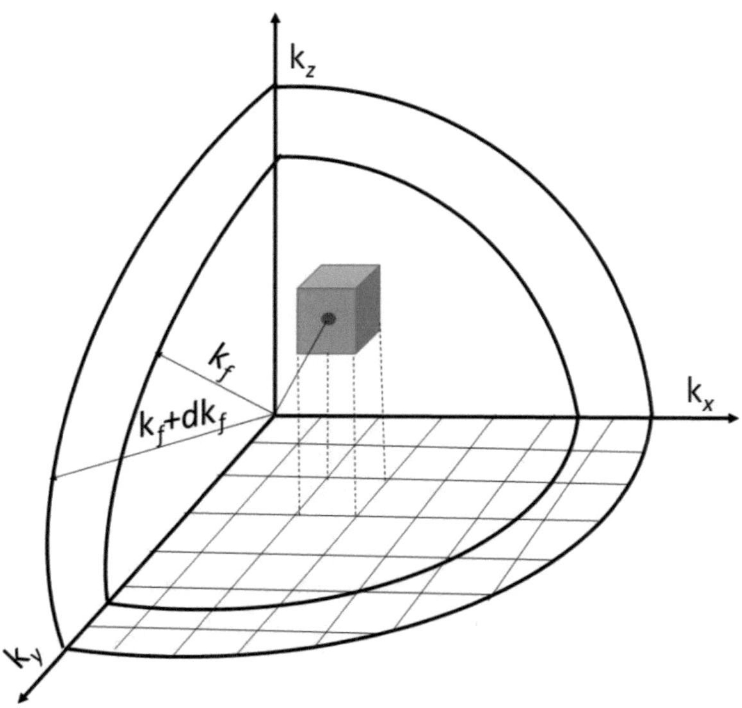

Figure 2.
The elemental volumes in the reciprocal space. Here the cube enclosing a lattice point represents the unit cell of size $V_{min} = (2\pi/L)^3$ (see text).

Therefore, one has

$$\left(\frac{d^2n}{d\Omega dE_f}\right)dE_f\Delta\Omega = k_f^2\left(\frac{L}{2\pi}\right)^3 dk_f d\Omega. \tag{16}$$

For photons, the link between energy and wavevector is fixed by the linear law $E_f = \hbar ck_f$, which can be differentiated to obtain $dE_f = \hbar cdk_f$, thus eventually getting

$$\left(\frac{d^2n}{d\Omega dE_f}\right)dE_f\Delta\Omega = \frac{k_f^2}{\hbar c}\left(\frac{L}{2\pi}\right)^3 dE_f d\Omega. \tag{17}$$

Therefore,

$$\frac{d^2n}{d\Omega dE_f} = \frac{L^3}{8\pi^2}\frac{k_f^2}{\hbar c}, \tag{18}$$

which, combined with Eq. (11), yields

$$\frac{d^2\sigma}{d\Omega dE_f} = \frac{L^3}{8\pi^2}\frac{k_f^2}{\hbar c}\frac{dP_{i\to f}}{dt}. \tag{19}$$

At this stage, the interaction term, that is, the squared matrix element appearing in the Fermi Golden Rule (Eq. (13)), can be made explicit by inserting in it the Thomson term in Eq. (9), while using the expression of the vector potential in Eq. (10), thus eventually obtaining

$$\sum_{j,m}\langle F|\, exp\,(-i\boldsymbol{Q}\cdot\boldsymbol{R}_j)|I\rangle\langle I|\, exp\,(i\boldsymbol{Q}\cdot\boldsymbol{R}_m)|F\rangle, \tag{20}$$

where the vector R_j is the position of the jth atom. The above formula embodies the momentum conservation law as it was derived assuming the identity $\hbar\boldsymbol{Q} = \hbar(\boldsymbol{k}_f - \boldsymbol{k}_i)$. Furthermore, when using Eq. (10), it was considered that $\omega(k)$ is equal to ck_i and ck_f in the initial and the final photon states, respectively. Combining all analytical steps illustrated above, one eventually obtains the following expression for the double differential cross-section:

$$\frac{\partial^2\sigma}{\partial\Omega\partial E_f} = r_0^2\frac{k_f}{k_i}\left(\hat{\varepsilon}_i\cdot\hat{\varepsilon}_f\right)^2 \times \sum_{F,I}P_I\left|\langle F|\sum_j exp\,(i\boldsymbol{Q}\cdot\boldsymbol{R}_j)|I\rangle\right|^2\delta(\hbar\omega + E_F - E_I). \tag{21}$$

Here, $\hbar\omega$ is the energy gained by the photons in the scattering process, while the δ-function term accounts for the energy conservation in the scattering process, as it ensures that $\hbar\omega = -(E_F - E_I)$ with $E_F - E_I$ being the energy gained by the sample. Notice that the cross-section defined above entails a sum over all states of the system, where the factor P_I represents the statistical population of the initial states of the sample.

5. From the adiabatic approximation to the dynamic structure factor

The right-hand side of Eq. (21) contains three independent factors, the integral term being the only one directly relating to the properties of the target sample. The latter can hardly be handled analytically, due to the complex interplay between electrons belonging to different atoms, which couples electronic and nuclear coordinates. However, it becomes treatable under the reasonable approximation that the centre of mass of the electronic cloud drifts following with no delay the slow nuclear motion. This assumption is customarily referred to as 'adiabatic', or Born-Oppenheimer, approximation [11]; its use justifies the factorization of the target system 'ket' as $|S\rangle = |S_n\rangle|S_e\rangle$, with nuclear and electronic states being labeled by the suffixes 'n' and 'e' respectively. The accuracy of this assumption ultimately owes to the substantially different nuclear and electronic masses and the correspondingly different timescales defining their dynamics. It holds validity when the energy exchange is smaller than all excitation energies of electrons in bound core states, which includes all cases of practical interest for this book. With $|S_e\rangle$ being unaffected by the scattering process, the difference between the initial $|I\rangle = |I_n\rangle|I_e\rangle$ and the final $|F\rangle = |F_n\rangle|F_e\rangle$ states of the sample is uniquely due to excitations associated with atomic density fluctuations.

Within the validity of these assumptions, the double differential cross-section in Eq. (18) reduces to

$$\frac{\partial^2\sigma}{\partial\Omega\partial E_f} = r_0^2\frac{k_f}{k_i}\left(\hat{\varepsilon}_i\cdot\hat{\varepsilon}_f\right)^2\sum_{F_n,I_n}P_{I_n}\left|\langle F_n|\sum_j f_j(\mathbf{Q})\,exp\,(i\mathbf{Q}\cdot\mathbf{R}_j)|I_n\rangle\right|^2 \times \delta(\hbar\omega + E_{F_n} - E_{I_n}), \tag{22}$$

where E_{I_n} and E_{F_n} are the energies associated with the initial and final nuclear states respectively, and

$$f_j(\mathbf{Q}) = \langle F_e|\sum_{\alpha=1}^{Z} exp\,\left(i\mathbf{Q}\cdot\mathbf{r}_\alpha^j\right)|I_e\rangle \tag{23}$$

is the form factor of the jth atom. Here r_α^j is the coordinate of the αth electron in the centre of mass frame of the jth atom, while $|I_e\rangle$ coincides with the ground state of the electronic wave function of a given atomic nucleus. In practice, $f(\mathbf{Q})$ can be approximated by the value calculated for a free atom, that is, in the perfect gas phase, as the electronic cloud distribution is essentially unchanged upon phase transition. The primary contribution to this factor comes from core electrons whose orbits are more tightly bound to the much more massive atomic nucleus.

If a single atomic species is present in the sample, all atoms have the same form factor, that is, $f_j(\mathbf{Q}) \equiv f(\mathbf{Q})$; this further simplifies the expression of the double differential cross-section

$$\frac{\partial^2\sigma}{\partial\Omega\partial E_f} = K\sum_{F_n,I_n}P_{I_n}\left|\langle F_n|\sum_m exp\,(i\mathbf{Q}\cdot\mathbf{R}_m)|I_n\rangle\right|^2 \delta(\hbar\omega + E_{F_n} - E_{I_n}) \tag{24}$$

where $K = r_0^2\frac{k_f}{k_i}\left(\acute{\varepsilon}_i\cdot\acute{\varepsilon}_f\right)^2|f(\mathbf{Q})|^2$.

The above expression can be cast in a more compact form after the few additional manipulations as the use of an integral representation of the δ-function of energy, the Heisenberg representation of a time-dependent operator and the completeness of the final eigenstate.

In its initial state, the sample is usually a many-atoms system at equilibrium, and the sum over its initial state can be computed as an ordinary equilibrium average, which leads to the following identity:

$$\sum_{I_n} P_{I_n} \langle I_n | \exp\left[-i\boldsymbol{Q}\cdot\boldsymbol{R}_k(0)\right] \exp\left[i\boldsymbol{Q}\cdot\boldsymbol{R}_j(t)\right] | I_n \rangle = \sum_{k,j} \langle \exp\left[-i\boldsymbol{Q}\cdot\boldsymbol{R}_k(0)\right] \exp\left[i\boldsymbol{Q}\cdot\boldsymbol{R}_j(t)\right] \rangle, \tag{25}$$

where as usual, the angle brackets $\langle \ldots \rangle$ denote the thermal average on the system at equilibrium. The expression above is the time correlation function of the variable $\sum_j exp\left[-i\boldsymbol{Q}\cdot\boldsymbol{R}_j(t)\right]$, which involves the pair composed by the jth and kth atoms. The physical meaning of this variable will be discussed in the next section in further detail.

In summary, as a result of all manipulations mentioned above, the double differential cross-section in Eq. (24) eventually reduces to

$$\frac{\partial^2\sigma}{\partial\Omega\partial E_f} = \frac{K}{2\pi\hbar}\int_{-\infty}^{\infty} dt \langle \sum_{j=1}^{N}\sum_{k=1}^{N} \exp\left\{i\boldsymbol{Q}\cdot\left[\boldsymbol{R}_j(t) - \boldsymbol{R}_k\right]\right\}\rangle \exp\left(-i\omega t\right) \tag{26}$$

where $\boldsymbol{R}_k$ is the shorthand notation for $\boldsymbol{R}_k(0)$.

6. Introducing a key stochastic variable: the microscopic density fluctuation

The expression between angle brackets is the equilibrium autocorrelation function of the dynamic variable $n(\boldsymbol{Q},t) = \sum_j \exp\left[i\boldsymbol{Q}\cdot\boldsymbol{R}_j(t)\right]$, which involves the positions of the generic kth and kth atom pair, that is, $\boldsymbol{R}_k(t)$ and $\boldsymbol{R}_j(0)$ respectively, evaluated at different times. The variable $n(\boldsymbol{Q},t)$ is the Fourier transform of the microscopic number density of the system, which, for a system of N atoms is defined as:

$$n(\boldsymbol{r},t) = \sum_{j=1}^{N} \delta\left[\boldsymbol{r} - \boldsymbol{R}_j(t)\right]. \tag{27}$$

The interpretation of this function as a microscopic density is perhaps more evident as one considers its average value over the whole sample volume:

$$n = 1/V\int_V d\boldsymbol{r} n(\boldsymbol{r},t) = 1/V\int_V d\boldsymbol{r}\sum_{j=1}^{N}\delta\left[\boldsymbol{r} - \boldsymbol{R}_j(t)\right] = N/V, \tag{28}$$

which is consistent with the macroscopic definition of number density. Notice that the δ-function is an extremely irregular discontinuous profile, which however adequately accounts for the atomistic, character of the system.

In the reciprocal space, one deals with the Fourier transform of the microscopic density, namely

$$n(\boldsymbol{Q},t) = \int_V d\boldsymbol{r}\left\{\sum_{j=1}^{N}\delta\left[\boldsymbol{r} - \boldsymbol{R}_j(t)\right]\right\} exp\left(i\boldsymbol{Q}\cdot\boldsymbol{r}\right) = \sum_{j=1}^{N} exp\left[i\boldsymbol{Q}\cdot\boldsymbol{R}_j(t)\right]. \tag{29}$$

Furthermore, since scattering phenomena arise from inhomogeneities or fluctuation from equilibrium, we are here mainly interested in the microscopic density fluctuation:

$$\delta n(\boldsymbol{r},t) = \sum_{j=1}^{N}\delta\left[\boldsymbol{r} - \boldsymbol{R}_j(t)\right] - n, \tag{30}$$

Again, the variable of direct pertinence for a spectroscopic measurement is instead the Fourier transform of such a fluctuation:

$$\delta n(\mathbf{Q},t) = \sum_{j=1}^{N} exp\left[i\mathbf{Q}\cdot\mathbf{R}_j(t)\right] - n\delta(\mathbf{Q}); \tag{31}$$

in which it was considered that the Fourier transform of a constant function is a δ-function.

We can now introduce the intermediate scattering function as the space Fourier transform of the correlation function between density fluctuations:

$$F(\mathbf{Q},t) = \frac{1}{N}\int_V d\mathbf{r}\langle\delta n(\mathbf{r},t)\delta n(\mathbf{r},0)\rangle\, exp\,(i\mathbf{Q}\cdot\mathbf{r}) \tag{32}$$

and its one-sided time Fourier transform

$$S(\mathbf{Q},\omega) = \frac{1}{2\pi\hbar}\int_0^{\infty} dt\, F(\mathbf{Q},t)\, exp\,(-i\omega t), \tag{33}$$

which is customarily referred to as the spectrum of density fluctuations, or the dynamic structure factor of the system.

7. The double differential cross-section and the dynamic structure factor

Given the dynamic variables introduced in the previous section, it can be readily verified that Eq. (26) can be cast in the more compact form:

$$\frac{d^2\sigma}{d\Omega dE_f} = N\frac{r_0^2}{\hbar}\left(\frac{k_f}{k_i}\right)\left(\hat{\varepsilon}_i\cdot\hat{\varepsilon}_f\right)^2|f(Q)|^2 S_n(Q,\omega), \tag{34}$$

where

$$S_n(Q,\omega) = \frac{1}{2\pi\hbar N}\int_{-\infty}^{+\infty} dt\langle n(\mathbf{Q},0)n(\mathbf{Q},t)\rangle\, exp\,(-i\omega t) \tag{35}$$

is the spectrum associated with the dynamic variable $n(\mathbf{Q},t)$. Notice that for a homogeneous and isotropic system such as a liquid, such a variable does not depend on the direction of the exchanged wavevector, but uniquely on its amplitude $Q = |\mathbf{Q}|$.

Let us discuss here how the spectrum in Eq. (35) relates to the variable density fluctuations as defined by Eq. (30). By definition, the spectrum of such a variable is the Fourier transform of the autocorrelation function. Explicitly

$$\begin{aligned} S_{\delta n}(Q,\omega) &= \frac{1}{2\pi\hbar N}\int_{-\infty}^{\infty} dt\langle\delta n(\mathbf{Q},t)\delta n(\mathbf{Q},0)\rangle\, exp\,(-i\omega t) \\ &= \frac{1}{2\pi\hbar N}\int_{-\infty}^{\infty} dt\langle n(\mathbf{Q},t)n(\mathbf{Q},0)\rangle\, exp\,(-i\omega t) + C\delta(\omega)\delta(\mathbf{Q}), \end{aligned} \tag{36}$$

with $C = n^2/\hbar N$. At this stage, one can define the spectrum of the microscopic density as

$$S_n(Q,\omega) = S_{\delta n}(Q,\omega) + n^2\delta(\omega)\delta(\mathbf{Q}). \tag{37}$$

It appears that the spectra of either $n(\mathbf{Q},t)$ or $\delta n(\mathbf{Q},t)$, which are labeled by the respective indexes n and δn, differ by a term proportional to the product $\delta(\omega)\delta(Q)$.

This term accounts for the forward transmitted elastic scattering, which is of no relevance for a scattering experiment as it describes the signal from photons that have exchanged no energy or momentum with the target sample.

In practice, such a signal is never detected by scattering measurements, as it does not convey insight into non-trivial samples properties; furthermore, it fully overlaps with the forward transmitted beam, which is often so intense to burn or damage detectors. For these reasons, IXS measurements are always performed at finite scattering angles, where one has

$$S_n(Q,\omega) = S_{\delta n}(Q,\omega) \equiv S(Q,\omega), \tag{38}$$

with

$$S(Q,\omega) = \frac{1}{2\pi N}\int_{-\infty}^{+\infty} dt \langle \delta n(\mathbf{Q},0)\delta n(\mathbf{Q},t)\rangle \, exp\,(-i\omega t), \tag{39}$$

As discussed, the identity above entails the replacement of the microscopic density $n(\mathbf{Q},t)$ with its fluctuation from equilibrium $\delta n(Q,t)$

$$\frac{d^2\sigma}{d\Omega dE_f} = KS(Q,\omega), \tag{40}$$

where $K = N\left(r_0^2/h\right)\left(k_f/k_i\right)\left(\hat{\varepsilon}_i \cdot \hat{\varepsilon}_f\right)^2 |f(Q)|^2$.

This expression of the cross-section above has been derived assuming a target sample composed by N identical atoms and within the Born-Oppenheimer approximation. When different atomic species are present in the sample, within the validity of the Born-Oppenheimer approximation, the derivation of the scattering cross-section is similar, provided the system is isotropic, that is, invariant under rotations, and a weak coupling exists between molecular rotations and centre of mass movements. The 'effective' form factor, in this case, results from the average value of the form factors of different atoms in the molecule. The general case, of a system composed of molecules with a pronounced anisotropy, that is, a markedly non-spherical shape, makes the computation of the cross-section slightly more complicated.

A more detailed treatment of this problem within the hypothesis of random molecular orientations and weak coupling between orientational and translational degrees of freedom leads to the conclusion that the spectrum splits into a coherent and an incoherent component. Consequently, the cross-section can be cast in the following general form:

$$\frac{\partial^2\sigma}{\partial\Omega\partial\omega} = A\left\{\langle F^2(Q)\rangle_\Omega S_C(Q,\omega) + \delta\langle F^2(Q)\rangle_\Omega S_I(Q,\omega)\right\} \tag{41}$$

where $\delta\langle F(Q)^2\rangle_\Omega = \langle F(Q)^2\rangle_\Omega - \langle F(Q)\rangle_\Omega^2$, where the suffix '$\Omega$' indicates an average over molecular orientations, while the suffixes 'I' and 'C' label the incoherent and coherent parts of the dynamic structure factor.

8. An estimate of the count rate

An estimate of the count rate achievable by an IXS measurement can be worked out starting from the expression of the total scattering cross-section, while assuming, for instance, a sample having an optimal thickness $t_s = 1/\mu$. The flux of scattered photons in the solid angle $\Delta\Omega$ and the energy interval ΔE_f is thus given by

$$d\dot{N} = \dot{N}_0 \, exp\,(-1) \frac{n_s}{\mu} \frac{d^2\sigma}{d\Omega dE_f} \Delta\Omega\Delta E_f. \tag{42}$$

At this stage, both members of the equation can be integrated in time, and the double integration over both solid angle and final energy must also be performed to obtain the total cross-section of the IXS scattering. In the low Q limit, where the atomic form factor $f(Q) \approx Z$ and that the approximation $(k_f/k_i)(\hat{\varepsilon}_i \cdot \hat{\varepsilon}_f)^2 = 1$, one has:

$$\frac{N}{N_0} \propto \frac{(Zr_0)^2 n_s}{\mu} = \frac{\sigma_C}{\sigma_A}, \tag{43}$$

where $\sigma_C = (Zr_0)^2$, while $\sigma_A = n_s/\mu$ is the absorption cross-section. An idea of the counting efficiency of IXS is provided by **Figure 3**, which displays the value of the σ_C/σ_A ratio for an incident X-ray beam having 10 keV energy, as a function of the atomic number. The abrupt increase of this parameter can be readily appreciated at the absorption above the K-edge, that is, above the binding energy of the innermost electron shell; these innermost electrons are those primarily interacting with the incident X-ray.

8.1 The signal measured by a real instrument

As a result of the previous treatment, it was demonstrated that the cross-section is proportional to $S(Q, \omega)$. However, such a treatment is entirely classical, insofar as

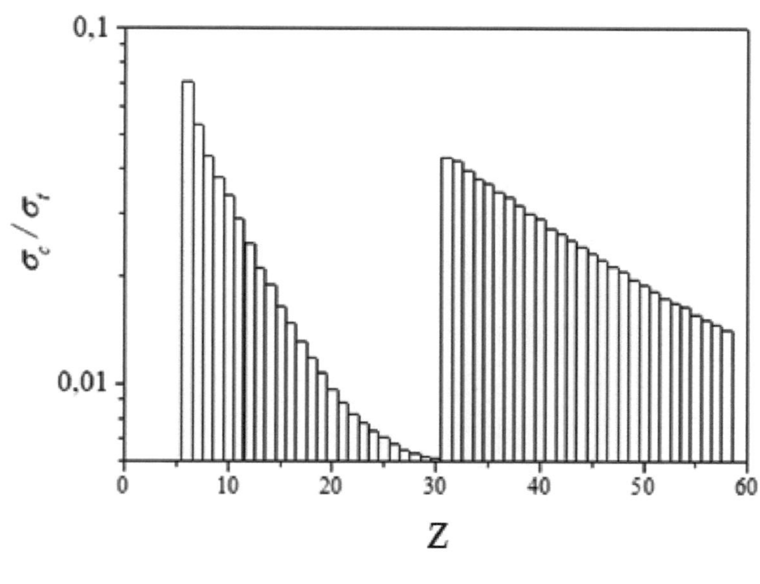

Figure 3.
The cross-sections ratio defined in Eq. (45) is reported as a function of the atomic number Z, for a 10 keV energy X-ray beam (courtesy of F. Sette).

all relevant observables are treated as commuting variables. Quantum effects are accounted for only through the so-called detailed balance principle, which takes into account the statistical population of the various $\hbar\omega$-states of the sample. These effects ultimately result in an asymmetry of the spectrum respect to its elastic, $\hbar\omega = 0$, position. The most popular recipe for handling them is to assume that the true spectrum $\tilde{S}(Q,\omega)$ can be obtained from the classic, symmetric, counterpart $S(Q,\omega)$ by adding a suitable frequency-dependent factor. Explicitly,

$$\tilde{S}(Q,\omega) = \frac{\hbar\omega}{k_B T}\left[\frac{1}{1 - exp\left(-\hbar\omega/k_B T\right)}\right] S(Q,\omega). \tag{44}$$

Still, the above formula does not capture two essential aspects of the measured scattering signal, as the contribution of the instrumental resolution and the spectral background. These are explicitly accounted for by using the following general expression for the intensity profile:

$$I = I(Q,\omega) = A\left[\tilde{S}(Q,\omega) \otimes R(\omega)\right] + B(\omega) \tag{45}$$

where A is an overall intensity factor, while the usually mildly frequency-dependent coefficient $B(\omega)$ accounts in principle for both the spectral background

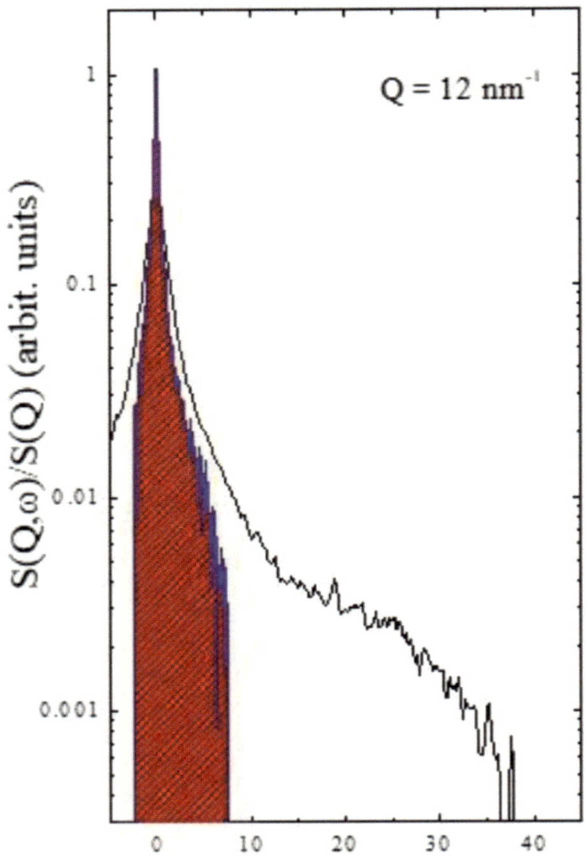

Figure 4.
The spectral line-shapes measured by IXS (black line) and INS (shadowed blue line) on a D_2O sample at ambient conditions. The spectra are reported after rescaling to the respective integrated intensities. Data are redrawn from ref. notice the remarkable difference in the explored ω-range. Data are redrawn from Ref. [13].

and the dark counts of a detector; when modeling the line-shape, sometimes the latter coefficient is assumed either constant or linearly dependent on ω.

The resolution profile represents the instrumental rendering of a spectral shape having zero energy width, that is, the $\delta(\omega)$–profile representing a perfectly elastic scattering. In a typical IXS measurement, such a resolution is estimated by measuring the scattering signal from an almost perfect elastic scatterer, often identified in a sample of plexiglas at the Q-position of the first sharp diffraction peak.

8.2 A practical example: a comparison between an IXS and an INS measurement

In general, the kinematic laws ruling the scattering process impose some limitations to the dynamic Q,ω region explorable by the measurement. These kinematic constraints are especially severe for inelastic neutron scattering, INS [2]. Although these limitations are irrelevant for IXS, the portion of the dynamic plane explored by this technique is still limited in the low-energy, or low-frequency, side by the finite instrumental energy resolution.

Figure 4 provides a clear example of how resolution and kinematic limitations differently affect IXS and INS. Indeed, the plot compares the spectra measured in a joint INS and IXS measurement on the same sample of heavy water [12], after normalization of the respective areas. The elastic peak in the INS spectrum has a spike-like shape. Such a sharp shape could be measured thanks to the 0.08 meV broad Gaussian resolution function. Which enables a superior definition of the spectral shape. However, this performance imposes an overall shrinkage of the spanned frequency range, which does not include the high-frequency shoulder in the IXS spectrum. On the other hand, the resolution of the IXS measurement is too coarse to enable a proper definition of the quasielastic portion of the scattering profile.

9. Conclusion

In conclusion, we illustrated the main analytical steps leading to a derivation of the inelastic X-ray scattering, IXS signal, and demonstrated its direct link with the terahertz spectrum of atomic density fluctuations.

Since its development in the mid-1990s, high-resolution IXS has rapidly transitioned to its mature age, nowadays representing an essential tool to characterize the terahertz dynamics of liquid and amorphous materials. Historically, the mainstream scientific interest of the IXS community was mostly limited to simple fluids and glass-forming materials. In recent years, such a focus has gradually shifted towards nanostructured metamaterials and biological systems. Since the high complexity of these systems often challenges a firm understanding of the measurement outcome, a firm theoretical modeling of the IXS signal from these highly heterogeneous systems would be highly beneficial.

Acknowledgements

This work used resources of the National Synchrotron Light Source II, a U.S. Department of Energy Office of Science User Facility operated for the DOE Office of Science by Brookhaven National Laboratory under Contract No. DE-SC0012704.

Author details

Alessandro Cunsolo
National Synchrotron Light Source-II, Brookhaven National Laboratory, Upton, NY, USA

*Address all correspondence to: acunsolo@bnl.gov

References

[1] Lovesey SW. Theory of Neutron Scattering from Condensed Matter. Oxford: Oxford University Press; 1984

[2] Squires GL. Introduction to the Theory of Thermal Neutron Scattering. Cambridge: Cambridge University Press; 1978

[3] Berne BJ, Pecora R. Dynamic Light Scattering. New York: Wiley; 1976

[4] Silver RN, Sokol PE. Momentum Distributions. New York: Springer Science; 1989

[5] Sinha SK. Theory of inelastic X-ray scattering from condensed matter. Journal of Physics. Condensed Matter. 2001;**13**:7511-7523. DOI: 10.1088/0953-8984/13/34/304

[6] Masciovecchio C, Bergmann U, Krisch M, Ruocco G, Sette F, Verbeni R. A perfect crystal X-ray analyzer with 1.5 meV energy resolution. Nuclear Instruments and Methods in Physics Research. 1996;**117**:339-340. DOI: 10.1016/0168-583X(96)00334-5

[7] Verbeni R, Sette F, Krisch MH, Bergmann U, Gorges B, Halcoussis C, et al. X-ray monochromator with 2×10^8 energy resolution. Journal of Synchrotron Radiation. 1996;**3**:62-64. DOI: 10.1107/S0909049595015883

[8] Scopigno T, Ruocco G, Sette F. Microscopic dynamics in liquid metals: The experimental point of view. Reviews of Modern Physics. 2005;**77**:881-933. DOI: 10.1103/RevModPhys.77.881

[9] Jackson JD. Classical Electrodynamics. New York: Wiley; 1999

[10] Born M, Wolf E. Principles of Optics: Electromagnetic Theory of Propagation, Interference and Diffraction of Light. Cambridge: Cambridge University Press; 1999

[11] Bransden BH, Joachain CJ. Physics of Atoms and Molecules. Harlow: Pearson Education; 2003

[12] Balucani U, Zoppi M. Dynamics of the Liquid State. Oxford: Oxford University Press; 1994

[13] Cunsolo A, Kodituwakku CN, Bencivenga F, Frontzek M, Leu BM, Said AH. Transverse dynamics of water across the melting point: A parallel neutron and X-ray inelastic scattering study. Physical Review B. 2012;**85**:174305. DOI: 10.1103/PhysRevB.85.174305

Chapter 2

Bayesian Approach for X-Ray and Neutron Scattering Spectroscopy

Alessio De Francesco, Alessandro Cunsolo and Luisa Scaccia

Abstract

The rapidly improving performance of inelastic scattering instruments has prompted tremendous advances in our knowledge of the high-frequency dynamics of disordered systems, yet also imposing new demands to the data analysis and interpretation. This ongoing effort is likely to reach soon an impasse, unless new protocols are developed in the data modeling. This need stems from the increasingly detailed information sought for in typical line shape measurements, which often touches or crosses the boundaries imposed by the limited experimental accuracy. Given this scenario, the risk of a bias and an over-parametrized data modeling represents a concrete threat for further advances in the field. Being aware of the severity of the problem, we illustrate here the new hopes brought in this area by Bayesian inference methods. Making reference to recent literature results, we demonstrate the superior ability of these methods in providing a probabilistic and evidence-based modeling of experimental data. Most importantly, this approach can enable hypothesis test involving competitive line shape models and is intrinsically equipped with natural antidotes against the risk of over-parametrization as it naturally enforces the Occam maximum parsimony principle, which favors intrinsically simple models over overly complex ones.

Keywords: inelastic X-ray scattering, inelastic neutron scattering, Bayes analysis, MCMC methods, model choice

1. Introduction

In the last decade, a large amount of inelastic neutron and X-ray scattering measurements focused on the study of the collective atomic dynamics of disordered system [1–5]. Although, across the years, the analysis of the line shape reported in these measurements seldom benefited from the support of a Bayesian inference analysis, the need of this statistical tool is becoming increasingly urgent. As a general premise, it is worth stressing that a scattering measurement somehow resembles a microscope pointed on the dynamics, whose "focus" can be adjusted by suitable choice of the momentum $\hbar Q$ and the energy $E = h\omega$ exchanged between the particle beam and the target sample in the scattering event, where $\hbar$ is the reduced Planck constant, Q is the wave vector transfer, and ω is the angular frequency. Specifically, upon increasing Q the probe "perceives" the response of the system as an average over smaller and smaller distances $\approx 2\pi/Q$ and over times $\approx 2\pi/\omega$ including a decreasing number of elementary microscopic events, e.g., mutual interatomic collisions. The observable accessed by these spectroscopic methods is the spectrum associated to density fluctuations, either spontaneous or scattering induced. When quasi-macroscopic distances are probed, i.e., in the $Q \to 0$ limit, the detail of atomic structure is lost, and the target sample is perceived as a continuum medium, whose dynamic behavior is recorded as an average over many elementary events [6]. Being the mass conserved over macroscopic scales, at these distances the liquid density tends to become a constant of motion, i.e., a time-invariant. For this reason, quasi-macroscopic density fluctuations relax very slowly to equilibrium, and collective density oscillations are correspondingly very long-living. The typical spectral signature of this so-called hydrodynamic behavior is a very sharp triplet reflecting the quasi-conserved nature of hydrodynamic density fluctuations. A striking example of such sharp triplet shape is provided in panel A of **Figure 1**, where the low $-(Q, \omega)$ spectrum on liquid argon at the triple point is reported as measured by Brillouin visible light scattering (BVLS) [7].

One could guess that such a sharp spectral shape does not leave any room for interpretative doubts, also considering that the limiting hydrodynamic spectral profile is exactly known as analytically treatable starting from the application of mass, momentum, and energy conservation laws. Although these statements appear partly true, the very concept of "interpretative doubt" sounds grossly ill-defined before spelling out explicitly the accuracy required to the interpretation one alludes to. Despite its pioneering nature, the quality of the measurements in panel A seems certainly adequate for a precise determination of the side-peak position, probably not much so for a detailed analysis of the spectral tails, which are dominated by the slowly decaying resolution wings. Nonetheless such a shape might still appear a more encouraging candidate for a line shape analysis than its counterpart reported in panel B of **Figure 1** which is featured by broad and loosely resolved spectral features, besides a definitely poorer count statistics. Given that the latter result is fairly prototypical of terahertz spectroscopic measurements on simple disordered systems, one might wonder why, thus far, the analysis of these measurements failed to benefit from Bayesian inference methods as routine line shape analysis tools. Aside of hardly justifiable initial mistrusts, a likely explanation is that only recently these spectroscopic techniques transitioned to a mature age in which the very detection of collective modes in amorphous systems can no longer be considered a discovery in itself, and detailed analyses of the spectral shape are more and more common and required. Again, the take-on message of this course of events is that

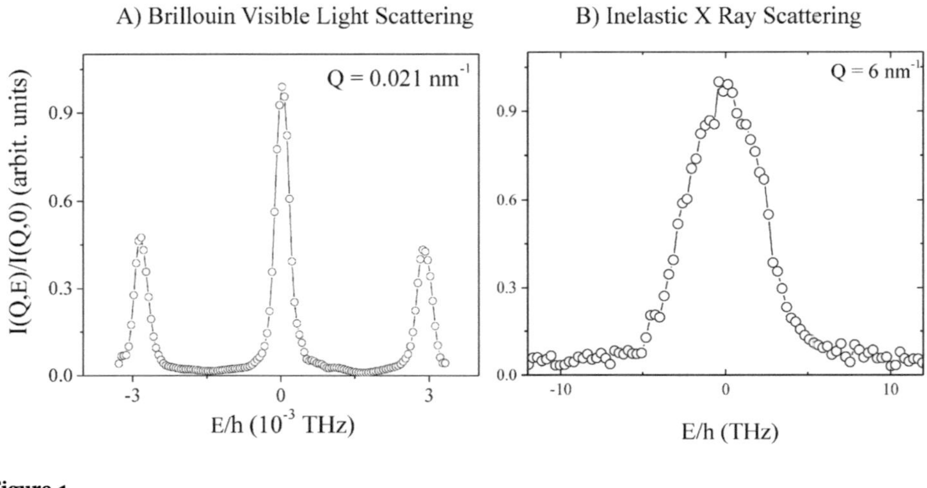

Figure 1.
Panel A: The Brillouin light scattering spectral intensity, $I(Q, E)$, measured in liquid argon at the triple point, redrawn from Ref. [7]. Panel B: The inelastic X-ray scattering spectrum of another noble gas: neon at ambient temperature and 0.3 GPa pressure [8]. Spectral profiles are normalized to their maxima.

the pivotal issue is the adequacy of a given measurement to provide the sought for information, rather than the quality of the measurement in itself. The unbalance between an unavoidably limited experimental performance and the rapidly increasing interpretative needs dramatically enhances the risk of "good faith over-interpretations" representing a lethal threat for the progress of knowledge.

Listen, the data talk! Every time we need to proceed with a data analysis, we could be induced or even tempted, on the basis of our prior knowledge or intuition, to somehow suggest the data what they should tell us about the properties of the system we are investigating. Being driven by acquired knowledge is not necessarily a wrong attitude, it is actually a natural demeanor which effectively drives the cognitive process and the progress of knowledge. However it could become deceiving if we do not have well-consolidated insight about the system under investigation and the observed data are not accurate enough or barely informative. In such cases, in fact, it is highly probable that we just adapt a model to the data, which fits them as well as many other possible models, with the only advantage to deliver results and solutions we feel more at ease with, as they confirm our prior beliefs. This model, of course, can be really plausible and reasonably pondered, and the solution adopted can accidentally be the right one; however, it would be desirable a robust method to quantify how much we can trust such a solution, either in itself or in comparison with alternative ones. We surely want to avoid an aprioristic reliance in a model, which might coerce data to confirm certain results preventing them from providing new insights on the investigated system.

When dealing with neutron or X-ray scattering, the statistical accuracy of spectral acquisition is the primary concern. For the most varied reasons, e.g., relating to the scattering properties of the sample, the integration time, or the count rate of the measurement, the achieved count statistics may either be adequate for a rigorous data analysis or, as often happens, not as good as we would like it to be. In the latter case, the experimental data might not be accurate enough to tell us everything about the physical problem under scrutiny. They could tell us something, but not everything! This is why we need a solid inferential method capable of extracting the maximum amount of information from the data acquired and possibly providing us with a quantitative probabilistic evaluation of the different models that are compatible with the data at hand. Especially when nothing or very little is known about a specific sample or system, the point is, given the observed data, how plausible is a specific model? What is the precision of the conclusions drawn from this model? Are there other possible interpretations of the data at hand? To what extent are different models and interpretations supported by the observed data?

A Bayesian inferential approach provides answers to all these questions on a probabilistic basis, along with a sound criterion to integrate any prior knowledge in the process of data analysis. Bayesian inference, in fact, recognizes the importance of including prior knowledge in the analysis. When we do have well-established prior knowledge about a sample property or a general law a physical phenomenon must comply with, it would be insane and pointless not to use this information. Such a prior knowledge, in fact, can protect us from the risk of making mistakes in the description of experimental data, hence in their interpretation. In the Bayesian framework, prior knowledge takes the form of probability statements so that different probabilities, ranging from zero to one, can be attributed to competitive explanations of the data. In this way, less probable explanations are not excluded a priori but simply given a smaller prior probability. The a priori probability of different explanations is then updated, through the Bayes theorem, based on the new information provided by the data. The results of this analysis, thus, assume the form of posterior probabilities. On this basis, one can easily establish which model is most supported by both data and prior knowledge, what are the posterior probabilities of alternative models and those of their parameters, and which provides a ground to appreciate the precision of their estimates. In addition, Bayesian methods naturally embody the Occam's razor principle, thus favoring simpler models over unnecessarily complex ones. Last but not least, Bayesian estimation algorithms are generally less affected by the presence of local optima in the parameter space and are not sensitive to the starting values used to initialize the estimation process.

The aim of this chapter is to illustrate how Bayesian inference can be used in X-ray and neutron scattering applications. The Bayesian approach proposed here is implemented through an estimation algorithm, which makes use of Markov chains Monte Carlo (MCMC) methods [9, 10] integrated, where appropriate, with a reversible jump (RJ) extension [11]. This Bayesian method has been already successfully applied in a series of Brillouin inelastic neutron scattering works [12], as well as inelastic X-ray scattering ones [13, 14] and, very recently, in the description of the time correlation function decay in the time domain as measured by spin echo neutron scattering [15, 16]. The rest of the work is organized as follows: Section 2 provides a motivating example; Section 3 revises the Bayes theorem and discusses its different components, as well as some advantages inherent in the Bayesian method; Section 4 applies the Bayesian inference to X-ray and neutron scattering spectroscopy with special emphasis on model choice, parameter estimation, and results interpretation.

2. An example: searching for differences

Depending on the problem at hand, our approach to data analysis can be very different. Imagine that we want, as a toy or teaching example, to measure either the neutron or the X-ray scattering spectrum from a system whose spectrum is well-known and its interpretation unanimously agreed. For instance, we aim at extracting the phonon dispersion curve from the thoroughly measured spectral density $S(Q, E)$ of a given sample. In our replica of past measurements, it is possible that the proper discernment of the excitation lines is hampered by both the course instrumental resolution and the limited statistical accuracy. The poor quality of data could prevent us from easily identifying the spectral features (peaks, bumps, shoulders), already measured and characterized by others. For instance, it could be overly difficult to establish how many excitations are present in the spectra. Unless we deliberately refute the conclusions previously reached by other scientists, it is natural to enforce a line shape modeling well-established in the kinematic range spanned and to verify ex post if the resulting spectral features are consistent with those known from literature.

More often, we face a different problem, as we want to measure for the first time a certain system on which we might not have previous knowledge. Alternatively, we could have prior knowledge about that same system, yet in different thermodynamic or environmental conditions— for instance, a liquid either in bulk or confinement geometries—and possible effects of these peculiar conditions are under scrutiny. Changes could also be very small, and, since detecting them is the focus of our research, it is essential to take the most impartial and noninvasive approach. In this second situation, it would be desirable not to rely too heavily on previous results when choosing the model and to allow the measurement to reveal possible new features.

The two situations mentioned above notably differ in the amount of information available on the system before analyzing the data. In the first case, we have a complete knowledge of the system, while, in the second case, this knowledge is partial or even lacking at all. In this second situation, a traditional approach would consist in either best fitting a model we deem adequate for the data, e.g., well-assessed for the considered sample, albeit only in different thermodynamic or environmental conditions, or fitting competing models to establish the one best performing based on criteria agreed upon, e.g., the chi-square value. Following the first path, we hinge too much on a specific model and on previous partial knowledge, thus jeopardizing the chance of new findings. On the other hand, the second path would be less coercive at the cost of completely ignoring previous partial knowledge. In addition, the model chosen would be simply the one providing the best fit, but no assessment can be made on the plausibility of this or any other fitting model, based on the data measured. Conversely, a Bayesian approach to data analysis would, instead, allow to assign a different prior probability to the different models (accounting for the uncertainty of available information on the system) and, then, revise these probabilities in the light of the data to deliver the posterior probability of each model, conditional on the data at hand.

3. Bayesian inference

3.1 The Bayes theorem

The Bayes theorem stems from the theorem of compound probability and from the definition of conditional probability. If we consider two events A and B, the compound probability theorem states that the probability of the two events occurring simultaneously is given by:

$$P(A,B) = P(B|A)P(A) = P(A|B)\mathrm{P}(\mathrm{B}), \tag{1}$$

where $P(B|A)$ is the probability of observing B, once A has been observed. Obviously, if A and B are independent, so that the occurrence of one of them does not affect the probability of occurrence of the other one, the compound probability theorem reduces to:

$$P(A,B) = P(A)P(B). \tag{2}$$

From Eq. (1), we immediately get:

$$P(A|B) = \frac{P(B|A)P(A)}{P(B)} \tag{3}$$

which is nothing else than the Bayes theorem.

Let us now consider A as the ensemble of the parameters of a certain model (or class of models) we choose to describe experimental data. In a slightly different notation, let this ensemble be denoted, from now on, as the vector $\Theta = (\theta_1, \theta_2, \dots, \theta_m)$, where each vector component θ_m is a parameter. Notice that a component of Θ might also be associated to a parameter that designates a particular model among several proposed. Also, consider B as the entire set of experimental data. Let us indicate this dataset with the vector $y = (y_1, y_2, \dots, y_n)$, with n being the sample size. In this new notation, the Bayes theorem reads obviously as:

$$P(\Theta|y) = \frac{P(y|\Theta)P(\Theta)}{P(y)}, \tag{4}$$

where $P(\Theta|y)$ is the posterior distribution of the parameters given the observed data; $P(\Theta)$ is the prior distribution of the parameters before observing the data;

$P(y|\Theta)$ is the likelihood of the data, i.e., the probability of observing the data conditional on a certain parameter vector; and $P(y)$ is the marginal probability of the data, which plays the role of normalizing constant so that Eq. (4) has a unit integral over the variable Θ. The different elements of Eq. (4) are thoroughly discussed in the following sections.

3.2 The prior distribution

Let us consider the different elements of Eq. (4), starting with the prior distribution (or simply prior) $P(\Theta)$. This is the distribution function *elicited* for the parameters, given the information at our disposal *before* data collection. Using a slightly more redundant notation, the prior can be explicitly denoted as $P(\Theta|I)$, where I represents the a priori information. This prior probability includes all prior knowledge (or lack of it) we might have, and it can be more or less informative depending on the amount of information on the problem under investigation. Using the same explicit notation, the Bayes formula in Eq. (4) can be rewritten as:

$$P(\Theta|y, I) = \frac{P(y|\Theta, I)P(\Theta|I)}{P(y|I)}. \tag{5}$$

Just to make a few examples, it might be possible that a certain parameter θ included in the model is known, or either already measured or somehow evaluated independently, and its value is $\theta^{\star}$. In this case, we can assume that the parameter takes the specific value $\theta^{\star}$ with probability equal to one. Otherwise, if we want to be less coercive, we can adopt for the parameter a Gaussian prior centered on $\theta^{\star}$ and with a variance opportunely chosen to limit the parameter variability to a certain interval around $\theta^{\star}$. In this way, values closer to $\theta^{\star}$ will be given a higher a priori probability.

In other situations, the information available on the parameters might be more vague. For example, we might simply know that a certain parameter must be nonnegative or that it must range in a limited interval, as often the case of neutron scattering hampered by severe kinematic constraints. Nonnegative parameters can be a priori assumed to follow, for example, a truncated Gaussian or a gamma distribution, and, if no other information is available, the prior distribution will be adjusted to make allowance for a large parameter variability, reflecting the noninformative initial guess. Parameters having random or hardly quantifiable variations within limited windows can be assumed to approximately follow a uniform distribution over such a window. Also, whenever feasible, any mutual entanglement between parameters, as well as any selection, conservation, or sum rule, should be embodied in a usable distribution function complementing our prior knowledge I in the cognitive process.

Notice that, even if it is common practice to assume that the parameters are a priori independently distributed, correlation between them can be naturally induced by the data, through the combination of the likelihood and the prior. Parameters can be a posteriori correlated, even if they are a priori independent.

3.3 The likelihood function

The likelihood function is the joint probability of the observed data, conditional on the model adopted and its parameter values. Notice that for continuous data, the likelihood becomes a density of probability. Let $y = (y_1, y_2, \ldots y_n)$ be a sample of data. Each datum y_i can be portrayed as a particular realization of a random variable

Y_i distributed as $f(Y_i; \Theta)$. In fact, if we had to collect data again, even under the same experimental conditions, we would obtain a different sample of data. This means that, before collecting data, the i-th result is to be considered a random variable Y_i. Once the data have been collected, y_i is the particular realization observed of the random variable Y_i, and the sample $y = (y_1, y_2, \ldots y_n)$ is the particular realization observed of the multiple random variable $Y = (Y_1, Y_2, \ldots Y_n)$, whose components Y_i are independent and identically distributed as $f(Y_i; \Theta)$. Then, the joint (density of) probability of the observed data is the probability that, simultaneously, each variable Y_i takes the value y_i (or takes a value in the interval $[y_i, y_i + \Delta y_i]$), for $i \in 1, \ldots, n$, i.e., $f(y_1, y_2, \ldots y_n, \Theta)$. Given the independence of the variables Y_i, for $i \in 1, \ldots, n$, and using the compound probability theorem, we obtain:

$$f(y_1, y_2 \cdots y_n; \Theta) = f(y_1; \Theta) f(y_2; \Theta) \cdots f(y_n; \Theta). \tag{6}$$

The left side of Eq. (6) is the likelihood function for the observed sample $y = y_1, y_2, \cdots y_n$, which depends on the unknown parameter vector Θ. If we condition on a particular value of Θ, we can compute the probability (or density) of the observed sample, conditional on Θ, i.e., $P(y|\Theta)$ in Eq. (3).

To be more specific, we can consider spectroscopic data. The observable directly accessed by a spectroscopic measurement is the spectrum of the correlation function of density fluctuation, or dynamic structure factor $S(Q, E)$, which, in a scattering experiment, is a unique function of the energy, $E = \hbar\omega$, and the momentum, $\hbar Q$, exchanged between the probe particles and the target sample in the scattering process. One has:

$$y_i = S(Q, E_j) + \varepsilon_i, \tag{7}$$

where $S(Q, E)$ is the model used for the dynamic structure factor, depending on a vector of unknown parameters Θ, and $\varepsilon = (\varepsilon_1, \varepsilon_2, \cdots, \varepsilon_n)$ is a vector of random errors, here assumed to be independently and normally distributed, i.e., $\varepsilon_i \sim N(0, \sigma_i^2)$, for $i \in 1, \ldots, n$. Notice that assuming heteroscedastic errors, we are not imposing any restriction other than normality on the error term. The heteroscedastic model embeds the homoscedastic one, and since the parameters σ_i^2 are estimated from the data, it might reduce to it if the data were compatible with the homoscedasticity constraint.

Under the assumption above, the likelihood function is:

$$P(y|\Theta) = \prod_{i=1}^{n} \frac{1}{\sqrt{2\pi\sigma_i^2}} e^{-\frac{[y_i - S_i(Q,E_i)]^2}{2\sigma_i^2}} = cost \cdot e^{-\sum_{i=1}^{n} \frac{[y_i - S(Q,E_i)]^2}{2\sigma_i^2}} \tag{8}$$

Conditional on a certain value of the parameter vector Θ (which might also include the variance σ_i^2 of the error term), we can compute $S(Q, E_i)$ and, thus, $P(y|\Theta)$.

3.4 The posterior distribution and its normalizing constant

The term on the left-hand side of Eq. (3) is the joint posterior distribution of the model parameters, given prior knowledge and measured data, i.e., *after* data collection. It incorporates both prior knowledge and the information conveyed by the data, and Bayesian inference completely relies on it. In practice, prior knowledge about the investigated problem is modified by the data evidence (through the likelihood function) to provide the final posterior distribution (**Figure 2**). Estimates for a single parameter θ_k can be obtained by marginalizing, i.e., by integrating

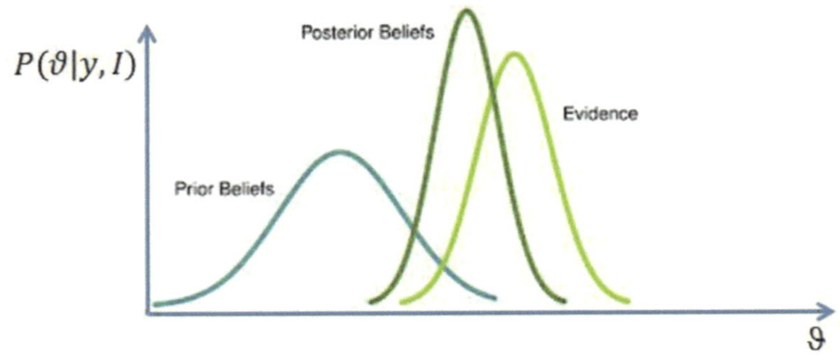

Figure 2.
Sketch of how the prior distribution and therefore our prior knowledge about a model parameter are changed by the data evidence.

(summing) the posterior over all the other parameters to get $P(\theta_k|y) = \int_{\Theta_{-k}} P(\Theta|y)d\Theta_{-k}$, where Θ_{-k} is the whole parameter vector except θ_k. Then, the mean of $P(\theta_k|y)$ is taken as a point estimate of θ_k, while the square root of its variance provides a measure of the estimation error. Also, the probability that the parameter θ_k belongs to a certain interval can be inferred from its marginal posterior.

The term in the denominator of Eq. (3):

$$P(y) = \int P(y|\Theta)P(\Theta)d\Theta \tag{9}$$

is generally called the marginal likelihood and represents the probability of observing the measured data $y_i (i = 1, \cdots n)$, averaged over all possible values of the model parameters. It represents the normalization constant for the posterior distribution, and it is required in the evaluation of $P(\Theta|y)$. However, in most cases, $P(y)$ does not have a closed analytical expression, as its determination would require the computation of high-dimensional integrals. Hence, the posterior distribution can only be obtained up to a normalizing constant, namely:

$$P(\Theta|y) \propto P(y|\Theta)P(\Theta). \tag{10}$$

For this reason, Bayesian inference usually needs to resort to MCMC methods to simulate the joint posterior distribution. MCMC algorithms, in fact, allow to draw values from distributions known up to a normalizing constant, as is often the case for $P(\Theta|y)$. Inference is then carried out on the basis of the simulated, rather than analytical, joint posterior distribution. More details on these methods will be given in Section 3.6 (see also Refs. [9, 10]).

To illustrate an interesting point, let us go back to the example considered before, in which we want to analyze spectroscopic data that can be modeled as in Eq. (7) and for which the likelihood is given in Eq. (8). Imagine to have no prior information at all on the parameters of the model so that the only sensible choice for the prior is a uniform distribution on the parameter space. Then, from Eqs. (8) and (10), it follows that:

$$P(\Theta|y) \propto \prod_{i=1}^{n} \frac{1}{\sqrt{2\pi\sigma_i^2}} \exp\left(-\frac{[y_i - S(Q, E_i)]^2}{2\sigma_i^2}\right) \propto \exp\left(-\sum_{i=1}^{n} \frac{[y_i - S(Q, E_i)]^2}{2\sigma_i^2}\right) \tag{11}$$

which implies that the posterior distribution is a multivariate Gaussian. As already mentioned, parameters can be estimated taking the mean of the posterior distribution, which, for a Gaussian distribution, corresponds to the median, mode, and maximum of the distribution. Therefore Bayesian parameter estimates are obtained as those values of Θ that maximize $\exp\left(-\Sigma_{i=1}^{n}\frac{[y_i - S(Q, E_i)]^2}{2\sigma_i^2}\right)$. This maximization is equivalent to the minimization of the $\chi^2 = \Sigma_{i=1}^{n}\frac{[y_i - S(Q, E_i)]^2}{2\sigma_i^2}$ function and thus provides the same estimates we would obtain through standard fitting procedures [13]. Therefore, whenever no prior information is available, which translates into a uniform prior, and a normal error distribution is assumed, the posterior distribution coincides up to a constant to the classical likelihood function, and Bayesian and classical estimates are equivalent. This result can be extended to the case of an informative prior, for which, again, Bayesian and traditional approaches provide asymptotically the same results. In particular, as sample size increases, the posterior distribution of the parameter vector approaches a multivariate normal distribution, which is independent of the prior distribution. These posterior asymptotic results [17] formalize the notion that the importance of the prior diminishes as n increases. Only when n is small, the prior choice is an important part of the specification of the model. In such situations it is essential that the prior truly reflects existing and well-documented information on the parameters so that its use can significantly improve the precision of the estimates.

Despite the asymptotic equivalence, sometimes parameters are much easier estimated in a Bayesian rather than in a frequentist perspective. Frequentist estimation, in fact, is generally based on least squares or maximum likelihood methods, and this might be a problem in the presence of local optima. If, for example, the starting values of the parameters, needed to initialize the optimization algorithm, are close to a local optimum, the algorithm might be trapped in this suboptimal solution. As a consequence, different starting values might determine different solutions and, thus, parameter estimates. The Bayesian estimate of a parameter, as stated before, is instead obtained as the mean of its posterior distribution, marginalized with respect to all other parameters. This estimation procedure does not involve any minimization or maximization, and, thus, the fitting algorithm does not risk to get trapped in local optima, and the results are independent from starting values used in the MCMC algorithm used to simulate the posterior distribution (see Section 3.6). It might happen, obviously, that the posterior of one or more parameters is bimodal or multimodal. The presence of different parameter regions with high posterior density might suggest that the data show some evidence in favor of a more complex model but not enough for this model to have the highest posterior probability. In this case, it is not reasonable to use the mean as a point estimate for the parameters, since it might fall in a low posterior density region, and the mode of the posterior distribution can be used in its place. In such situations of posterior multimodality, it is evident how the whole posterior distribution conveys a much richer information than the simple parameter estimate.

3.5 The Occam's razor principle

Even if Bayesian and classical analysis asymptotically give the same results, Bayesian results always have a probabilistic interpretation, and this is particularly relevant when we need to compare different models and determine, for instance, the number of spectral excitations (in the frequency domain) or the number of relaxations (in the time domain). In addition, the Bayesian method represents a natural implementation of the Occam's razor [18–20]: this principle is intrinsic to

Bayesian inference and is a simple consequence of the adoption of the Bayes theorem. In model choice problems, in fact, the posterior probabilities of the different models naturally penalize complex solutions with respect to simple ones, thus conforming to the parsimony principle.

To see this, consider Eq. (4), and imagine that the parameter vector also includes a model indicator parameter M_j, for $j = 1, \dots, k$. To make this more explicit, we can rewrite Eq. (4) as $P(\Theta, M_j|y) \propto P(y|\Theta, M_j)P(\Theta, M_j)$. Then, Bayesian model choice simply consists in choosing the model with the highest posterior probability $P(M_j|y) = \int_\Theta P(\Theta, M_j|y)d\Theta \propto \int_\Theta P(y|\Theta, M_j)P(\Theta, M_j)d\Theta = P(M_j)\int_\Theta P(y|\Theta, M_j)$ $P(\Theta|M_j)d\Theta = P(y|M_j)P(M_j)$. Thus, if the same a priori probability is attributed to the models, i.e., $P(M_1) = P(M_2) = \dots = P(M_k)$, the posterior probability $P(M_j|y)$ is simply proportional to the marginal likelihood:

$$P(y|M_j) = \int_\Theta P(y|\Theta, M_j)P(\Theta|M_j)d\Theta \tag{12}$$

Now, consider for simplicity just two possible models, the first one, denoted as M_1, more complex and characterized by a larger number of parameters and the second one, denoted as M_2, simpler and characterized by a smaller number of parameters. Clearly, the more complex model is able to generate a much wider range of possible datasets (i.e., for which the model would provide a reasonable fit) than the smaller model. Therefore, the marginal likelihood $P(y|M_1)$ is more dispersed than $P(y|M_2)$ (cf. Figure 28.3 of Ref. [20]). This implies that dataset in accordance with both M_1 and M_2 have $P(y|M_2) > P(y|M_1)$, while those in accordance with just the more complex model M_1 have $P(y|M_2) < P(y|M_1)$ (with $P(y|M_2) \cong 0$). If the two models are a priori given the same probability, for datasets in accordance with both models, the inequality $P(M_2|y) > P(M_1|y)$ holds for the posterior probabilities, determining the choice of the simplest model to represent the data.

3.6 Bayesian computation of model parameters

As already stated, Bayesian inference completely relies on the joint posterior distribution $P(\Theta|y)$. However, for a complex model, it is often impossible to compute this posterior distribution analytically, and the latter is only known up to a normalizing constant. The MCMC methods allow to draw values from distributions known up to a normalizing constant and, thus, to obtain the simulated joint posterior distribution. In practice, MCMC methods consist in constructing an ergodic Markov chain (**Figure 3**) with states Θ^m, $m = 1 \cdots M$, and stationary distribution corresponding to the joint posterior distribution. M is the number of states, i.e., the number of

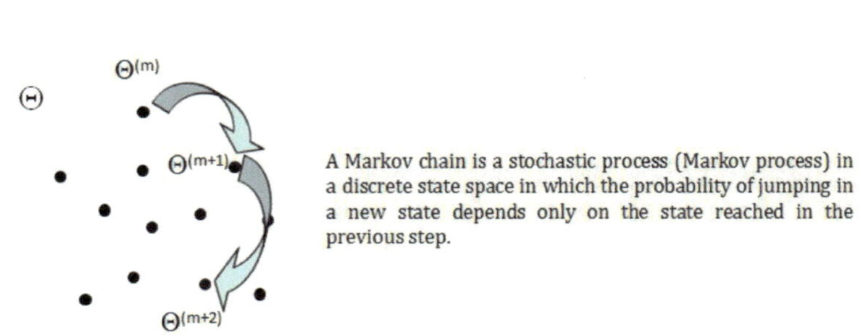

Figure 3.
Parameter updating. Θ *is the parameter vector.* $\Theta^{(m)}$ *is a particular set of parameter values in the parameter hyperspace.*

updating of the parameter values, and is generally called the number of sweeps of the MCMC algorithm. At each sweep of the algorithm, a new draw of Θ from its posterior is obtained updating all the parameters in turn, drawing each of them from its posterior distribution, conditional on the value of all the other parameters. If this posterior conditional distribution is known, the parameter is updated using a Gibbs sampling step, which simply draws the new value of the parameter from this known distribution. Otherwise, if this posterior conditional distribution is known only up to a normalizing constant, the parameter needs to be updated through a Metropolis-Hasting move [21]. This move is built as follows. Suppose that, at a given sweep of the algorithm, the current value of a certain parameter is θ. A new *candidate* value θ' can be drawn from an opportunely chosen *proposal* distribution $q(\cdot|\theta)$, which generally depends on the current value θ. The new value θ' is then accepted with a probability equal to $\min(1, R)$, where R is given by:

$$R = \frac{P(y|\Theta')}{P(y|\Theta)} \frac{P(\theta')}{P(\theta)} \frac{q(\theta|\theta')}{q(\theta|\theta)} \tag{13}$$

where Θ' is the whole parameter vector with the parameter θ replaced by the new value θ', $P(y|\Theta)$ is the likelihood, and finally $P(\theta)$ is the prior on that parameter. In other words, R is nothing else but the ratio between the joint posterior distribution calculated with the updated parameter values and the posterior distribution calculated with the current ones, multiplied by the ratio between the proposals, $q(\theta|\theta')/q(\theta'|\theta)$. The higher the posterior ratio, the larger R and hence the probability to move to the new parameter value. In practice, to decide whether or not a candidate value is accepted, a random number is drawn from a uniform distribution defined between 0 and 1 and compared with the calculated value for R. If the random number is less than R, the parameter is updated to the new value; otherwise the new value is rejected. The way the acceptation rule in Eq. (13) is built ensures that the resulting Markov chain has the joint posterior distribution $P(\Theta|y)$ as stationary distribution.

Concerning the proposal distribution, this should be chosen as a distribution from which it is easy to sample. It could be, for instance, a normal distribution centered on the current value of the parameter and with a certain variance which can be adjusted and used as a tuning parameter. This locution alludes to the circumstance that adjustments of this parameter can literally tune the step of the parameter updates. For a normal proposal distribution, a large variance allows the new value θ' to substantially change from the current value. However, if we already are in a high posterior distribution region for the parameter, values far from the current one will fall in low-density regions and are accepted with a very low probability. As a consequence, the algorithm will remain stuck on the same value of the parameter for a long time, causing an inefficient exploration of the parameter space. On the contrary, a small variance will constrain θ' to be close to θ. In this case, the new value has a high probability of being accepted, but the algorithm would move slowly and take a long time to reach convergence to the stationary distribution. The tuning parameters can be appropriately chosen so that the algorithm explores the parameter space efficiently. A rule of thumb states that this happens when the acceptance ratio for each parameter is about 30% [22].

When the parameter vector also includes a model indicator parameter, a further move needs to be considered to update this parameter and to allow the algorithm to explore different models. This move is a reversible jump [11] step, which is specifically designed to allow the Markov chain to move between states having different dimensions (since the dimension of the parameter space varies accordingly to the model considered).

As a final remark, consider that when the MCMC algorithm reaches convergence, after a so called "burn-in" period, the draws not only effectively represent samples from the joint posterior distribution but are also theoretically independent from the starting values of each parameter. Few examples about this point are shown in **Table 1** of Ref. [12]. Notice, however, that the time required to reach convergence might vary a lot depending on the data and the prior. For example, peaked unimodal posterior distributions (i.e., highly informative data) generally speed up convergence, as well as the availability of an important prior information, which reduces the size of the effectively accessible parameter space. On the contrary, the presence of many high posterior density regions can hinder and slow down convergence.

4. The Bayesian approach in neutron and X-ray scattering spectroscopy

4.1 Neutron and X-ray Brillouin scattering

One of the models commonly used to analyze either neutron or X-ray scattering data is the so-called damped harmonic oscillator (DHO) profile, which we report here below:

$$S(Q,E) = A_e(Q)\delta(E) + \frac{E}{k_B T}[n(E)+1]\left\{ L_{A_0,z_0}(Q,E) + \sum_{j=1}^{k} \frac{2}{\pi} \frac{A_j(Q)\Omega_j^2(Q)\Gamma_j(Q)}{\left[E^2 - \Omega_j^2(Q)\right]^2 + 4\left[\Gamma_j(Q)E\right]^2} \right\} \tag{14}$$

where $\delta(E)$ is the Dirac delta function describing the elastic response of the system modulated by an intensity factor $A_e(Q)$, $n(E) = \left(e^{E/k_B T} - 1\right)^{-1}$ is the Bose population factor expressing the detailed balance condition, and the term in curly brackets is the sum of a Lorentzian central contribution, characterized by the parameters A_0 and z_0, and the contribution of k pairs of peaks, the DHO doublets, symmetrically shifted from the elastic $(E = 0)$ position. The generic j-th DHO is characterized by its undamped oscillation frequency $\Omega_j(Q)$, damping $\Gamma_j(Q)$, and intensity factor $A_j(Q)$. The Lorentzian contribution, not necessarily present, accounts for the quasielastic response of the system. We have intentionally expressed the inelastic contribution as an indefinite sum of k terms, as the scattering signal from amorphous systems is often poorly structured and the number of inelastic modes contributing to it is often hard to guess. A series of concomitant factors, such as the instrument energy resolution, the limited statistical accuracy, and the intrinsically weak scattering signal, can make the line shape modeling not straightforward. In a Bayesian perspective, the number of inelastic features can be treated as a parameter to be estimated along with the other model parameters.

To fit the experimental data, the model in Eq. (14) needs to be convoluted with the instrument resolution, and it can conceivably sit on top of an unknown linear background. Overall, the final model used to approximate the measured line shape is given by:

$$\tilde{S}(Q,E) = R(Q,E) \otimes S(Q,E) + (\beta_0 + \beta_1 E). \tag{15}$$

where " $\otimes$ " represents the convolution operator. For neutron scattering, the instrument resolution function has often a Gaussian shape; thus the final model reads as:

$$\tilde{S}(Q,E) = \left[\frac{1}{\sqrt{2\pi}\zeta}\exp\left(-\frac{E^2}{2\zeta^2}\right)\right] \otimes S(Q,E) + (\beta_0 + \beta_1 E). \tag{16}$$

For IXS, Eqs. (14–16) are still formally valid although the instrument resolution function has usually a slightly more complex shape which appears in the convolution of Eq. (15) either as approximated by an analytical model or measured from the signal of an almost elastic scatterer; obviously, in the latter case, the convolution is computed numerically. The final model is further corrupted by an additive Gaussian noise, having a variance that, for instance, can be taken proportional to each data point. Thus, the experimental data points are given by:

$$y_i = \tilde{S}(Q,E_i) + \varepsilon(Q,E_i), \tag{17}$$

with

$$\varepsilon(Q,E_i) \overset{i.i.d}{\sim} \mathcal{N}\left(0, \sigma^2 \tilde{S}(Q,E)\right), \tag{18}$$

where σ^2 is the proportionality constant. Thus, the likelihood for model in Eq. (17) is simply given in Eq. (8), with $S(Q,E_i)$ replaced by $\tilde{S}(Q,E_i)$ defined in Eq. (16) and $\sigma_i^2 = \sigma^2 \tilde{S}(Q,E_i)$.

The whole parameter vector for the model in Eq. (17) is $\Theta = (k, A, \Omega, \Gamma, A_e, A_0, z_0, \beta_0, \beta_1, \sigma^2)$, with $A = (A_1, \cdots, A_k)$, $\Omega = (\Omega_1, \cdots, \Omega_k)$, and $\Gamma = (\Gamma_1, \cdots, \Gamma_k)$, so that the dimension of the parameter vector depends on the number of inelastic modes, k. In a Bayesian perspective, suitable priors need to be chosen for each component of Θ. For example, k can be safely assumed as uniformly distributed between 1 and a certain value $k_{\max}$ opportunely fixed so that all models are a priori given the same probability. All parameters only attaining nonnegative values such as $(A, \Omega, \Gamma, A_e, A_0, z_0$ and $\sigma^2)$ can, instead, be assumed distributed according to a Gamma distribution or a Gaussian distribution truncated in zero. Finally, β_0 and β_1 are assumed to follow a normal distribution, centered in zero and with a large variance, to keep the priors scarcely informative.

Bayesian inference is, then, based on the joint posterior of the whole parameter vector Θ. However, as mentioned, given the complexity of the model $\tilde{S}(Q,E_i)$ in Eq. (17), the normalizing constant in Eq. (9) cannot be analytically evaluated, and it is necessary to resort to MCMC methods to obtain a simulated joint posterior. Since the parameter space dimension depends on the number of inelastic modes, k, a RJ step needs to be added to allow the exploration of a parameter space of variable dimension. The updating of the parameter k can be implemented according to different types of moves, which, for instance, can enable either the creation (the birth) of a new component in Eq. (14) or the suppression (the death) of an existing one, i.e., the so-called birth-death moves; or they can promote the splitting of an existing component into two components or the combination of two existing components into one (split-combine move). These moves are described in Ref. [12]. In practice, at each step, the algorithm tries to jump to another value of k (from 1 to 2, from to 2 to 1 or 3, from 3 to 2 or 4, and so on). The new value of k is accepted with an acceptance probability that guarantees the convergence of the algorithm to the joint posterior distribution.

Once the convergence is attained, after a burn-in period, at each sweep $m = 1, \cdots, M$, the RJ MCMC algorithm draws a vector:

$$\left(k^{(m)}, A^{(m)}, \Omega^{(m)}, \Gamma^{(m)}, A_e^{(m)}, A_0^{(m)}, z_0^{(m)}, \beta_0^{(m)}, \beta_1^{(m)}, \sigma^{2(m)}\right), \tag{19}$$

from the joint posterior $P(\Theta|y)$. In practice, the output of the algorithm is a matrix of the form:

$$\begin{pmatrix} k^{(1)} & A^{(1)} & \Omega^{(1)} & \Gamma^{(1)} & A_e^{(1)} & A_0^{(1)} & z_0^{(1)} & \beta_0^{(1)} & \beta_1^{(1)} & \sigma^{2(1)} \\ k^{(2)} & A^{(2)} & \Omega^{(2)} & \Gamma^{(2)} & A_e^{(2)} & A_0^{(2)} & z_0^{(2)} & \beta_0^{(2)} & \beta_1^{(2)} & \sigma^{2(2)} \\ \cdots & \cdots & \cdots & \cdots & \cdots & \cdots & \cdots & \cdots & \cdots & \cdots \\ k^{(m)} & A^{(m)} & \Omega^{(m)} & \Gamma^{(m)} & A_e^{(m)} & A_0^{(m)} & z^{(m)} & \beta_0^{(m)} & \beta_1^{(m)} & \sigma^{2(m)} \\ \cdots & \cdots & \cdots & \cdots & \cdots & \cdots & \cdots & \cdots & \cdots & \cdots \\ k^{(M)} & A^{(M)} & \Omega^{(M)} & \Gamma^{(M)} & A_e^{(M)} & A_0^{(M)} & z_0^{(M)} & \beta_0^{(M)} & \beta_1^{(M)} & \sigma^{2(M)} \end{pmatrix} \tag{20}$$

where each row is a particular draw of the whole parameter vector Θ from its joint posterior $P(\Theta|y)$, while each column refers to a particular parameter and represents the whole simulated marginal posterior distribution for that parameter, independently from the values observed for all the other parameters. Model choice can, then, be accomplished considering the first column of the matrix in Eq. (20), that is, the simulated marginal posterior $P(k|y)$. This column contains a string of values for k (e.g., 1, 1, 1, 2, 2, 3, 2, 3, 4, 3,3,3, 2$_{\dots\ \dots}$4,5,4,3, 4, 3, 2$_{\dots}$). Therefore, the posterior probability that the number of modes is equal to a specific value $\ell, P(k = \ell|y)$, is given by the relative frequency of occurrence of the value ℓ in the strings, and the model chosen will be the one corresponding to the value of k with the highest occurrence.

Once a particular model with, let us say, $k = \ell$ inelastic modes has been chosen, the parameters of this model can be estimated conditionally on $k = \ell$. This means that we only need to consider a submatrix of the matrix in Eq. (20), made up of those rows for which the first column is equal to ℓ. Then, a certain parameter θ can be estimated taking the mean (or the mode) of the corresponding column of this sub-matrix, which represents the simulated posterior distribution for θ, conditionally on the model with ℓ modes and marginalized with respect to all the other parameters, i.e., $P(\theta|y, k = \ell)$.

In assessing convergence, a valuable tool is provided by trace plots, which show the sampled values of a parameter over the sweeps of the algorithm. Ideally, a trace plot should exhibit rapid up-and-down variation with no long-term trends or drifts. Imagining to break up this plot into a few horizontal sections, the trace within any section should not look much different from the trace in any other section. This indicates that the algorithm has converged to the posterior distribution of the parameters. Other convergence criteria can be found, for example, in Ref. [23]. **Figure 4** shows the trace plots of three DHO-mode frequencies ($\Omega_{1,2,3}$) the algorithm found, fitting a spectrum relative to IXS data on pure water recently measured (data not published) at room temperature and at a wave vector transfer $Q = 3\text{nm}^{-1}$, after the first 1000 (a), 10,000 (b), and 100,000 (c) sweeps. In plot (a), it can be seen how rapidly Ω_2 and Ω_3 reach their respective high-density regions, while Ω_1 has more problems in exploring the parameter space. Plot (b) shows that, after nearly 2000 sweeps, also Ω_1 finally starts oscillating around its mean, according to its posterior distribution. Plot (c) illustrates that a burn in of, for example, 10,000 sweeps is large enough to ensure convergence of the algorithm: the trace plots for the three parameters stabilize well before the end of the burn in period.

In **Figure 5**, we report an example of Bayesian analysis applied to neutron Brillouin scattering data from liquid gold [12] at different values of the momentum transfer Q. In this work, after a proper removal of spurious effects such as background, self-absorption, and multiple scattering, the data look indeed rather structured so that inferring the number of inelastic components seems rather obvious and the result confirms the findings of a previous work [24]. Estimates were obtained from 10^5 sweeps of the algorithm, after a burn in of 10^4 sweeps, and the running time for the algorithm was of approximately 5/10 minutes for each spectrum. We chose a precautionary large value for the burn-in, but convergence was normally achieved in a few hundreds of sweeps.

Even in this straightforward case, however, additional insights can be obtained from the posterior distributions delivered by the Bayesian inference. For example, in **Figure 6**, it can be noticed that, as the value of Q increases, the posterior probability of $k = 2$ also increases. This trend in the discrete distribution for k as a function of Q could possibly convey interesting insights on the actual onset of a second excitation or simply indicate a progressive degradation of the experimental data or, still, suggest that, as the damping becomes more and more effective, the determination of the number of inelastic features becomes more controversial.

To investigate these issues, one can look, for example, at the posterior distributions for the excitation frequency Ω, conditional to $k = 1$ and to $k = 2$, respectively (see **Figure 7** for an example on the same data of **Figure 5** for a Q value of 16 nm^{-1}). Considering the matrix in Eq. (20), as explained above, for $k = 1$, all the values in the column referring to Ω_1, and in correspondence with the rows for which $k = 1$, are draws from $P(\Omega_1|y, k = 1)$, and a histogram of these values can be used to visualize the marginal posterior distribution of Ω_1, conditional to $k = 1$. In the same way, for $k = 2$, all the values in the column referring to Ω_1, and in correspondence with the rows for which $k = 2$, are draws from $P(\Omega_1|y, k = 2)$, while those in the column referring to Ω_2 and in the same rows represent draws from $P(\Omega_2|y, k = 2)$. **Figure 7** illustrates, from left to right, the distributions $P(\Omega_1|y, k = 1)$, $P(\Omega_1|y, k = 2)$, and $P(\Omega_2|y, k = 2)$ at $Q = 16\text{nm}^{-1}$.

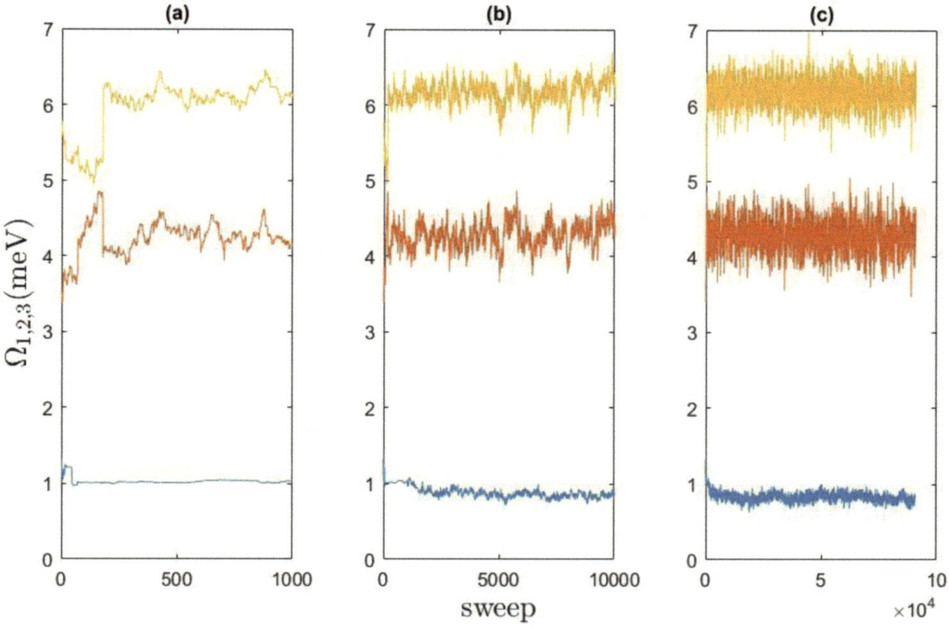

Figure 4.
Typical trace plots for the three DHO-mode frequencies as obtained after the first 1000 (a), 10000 (b) and 100000 (c) sweeps of the algorithm for IXS data on pure water at room temperature and at a wave vector transfer $Q = 3\text{nm}^{-1}$. The frequencies are indexed in increasing order with respect to their value.

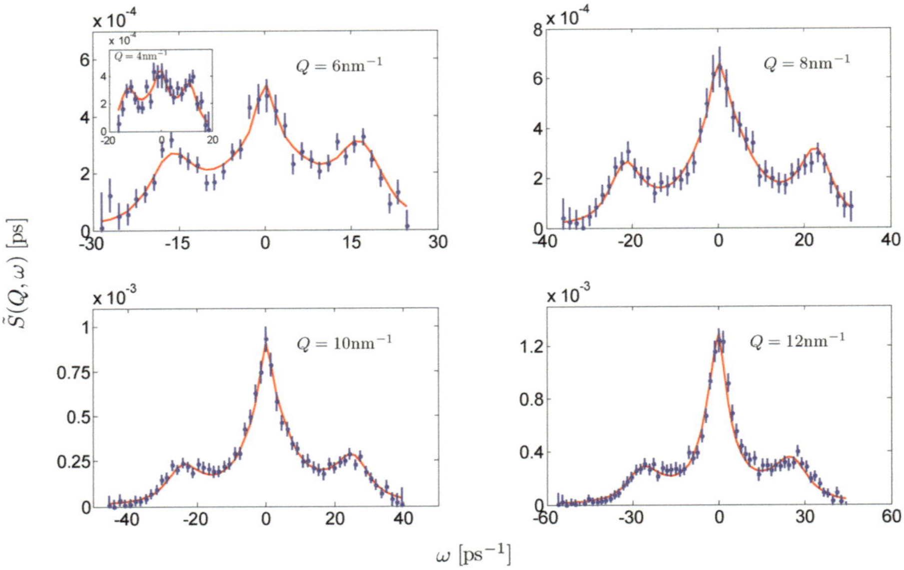

Figure 5.
Dynamic structure factor of liquid gold at five Q values measured on the Brillouin neutron spectrometer BRISP at ILL (Grenoble, France). The experimental data (blue dots) are broadened by the instrumental energy resolution. The RJ-MCMC best fit (red line) takes detailed-balance asymmetry and resolution into account. Reproduced from Ref. [12], Copyright (2016) of American Physical Society.

The shape of these posterior distributions provides a measure of the precision with which the parameter is estimated. For example, $P(\Omega_1|y, k = 1)$ is well-shaped, i.e., unimodal and approximately symmetric, yet quite dispersed. Its mean is equal to 23.8 meV, but there is a 95% probability that the value of Ω_1 lies in the large interva1 22.3–25.2 meV. This large interval tells us that many different values of Ω_1 are compatible with the data, signifying that the inelastic mode at $Q = 16$ nm^{-1} is largely damped—as confirmed also by the large Γ_1 value (= 7.5meV)—and less defined, which reveals the large uncertainty in the estimation of the undamped oscillation frequency of the *DHO* excitation. If we now look at the posteriors for $P(\Omega_1|y, k = 2)$ and $P(\Omega_2|y, k = 2)$, we can see that these are much worse shaped than $P(\Omega_1|y, k = 1)$, with unreasonably large or small values having nonvanishing probability. Their mean are, respectively, 17.6 and 25.5 meV and are outside the probability interval obtained for Ω_1, when $k = 1$. Therefore, based on these findings, the Q-evolution of the posterior probability of k seems to simply reveal the increasingly elusive discernment of distinct inelastic features as their damping, or broadening, increases. In practice, at the highest Q explored (16 nm^{-1}), the oscillation mode becomes so highly damped that it can be fitted equally well either by two distinct DHO peaks or by a (broader) single one in the middle of the two. At this stage, the Occam's razor comes into play, naturally integrated in the Bayesian model choice, which ultimately privileges the model with only one DHO, as it involves fewer free parameters. Imagine, instead, that $P(\Omega_1|y, k = 1)$ were bimodal, with the two modes corresponding to the single modes of $P(\Omega_1|y, k = 2)$ and $P(\Omega_2|y, k = 2)$, respectively, as observed, for instance, in Ref. [25]. In this case, the bimodality of $P(\Omega_1|y, k = 1)$ would have provided stronger support to the actual presence of two DHOs, thus suggesting that the finding $P(k = 1|y) > P(k = 2|y)$ only stemmed from the scarcity of data. Should this have been the case, additional observations would have probably led to privilege a more complex model.

Data discussed in Ref. [25] provide another example of the efficacy of Bayesian inference in enforcing the parsimony principle. Specifically, we refer to the case of

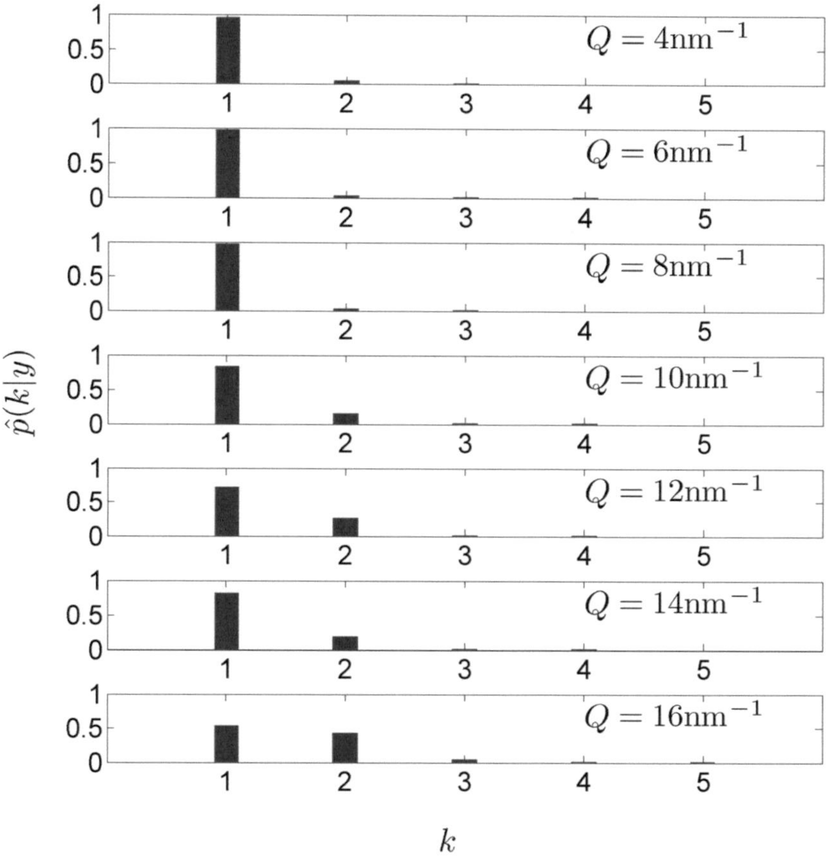

Figure 6.
*Posterior probability for the number of modes k at different values of Q for the spectra of **Figure** 5. Reproduced from Ref. [12], Copyright (2016) of American Physical Society.*

an IXS measurement from a suspension of gold nanoparticles in water which has been analyzed with a model similar to the one in Eq. (14), yet with the DHO terms replaced by Dirac delta functions, due to the extremely narrow width of the measured excitations. For all Q's explored, the posterior distributions for the number of inelastic modes have a maximum (**Figure 8**), which is smaller than $k_{\max}$. In particular, we can also observe that the most probable number of modes and the related probability change from one dataset to the other; this partly reflects the physics of the phenomenon under study but also drawbacks of the modeling, such as the limited count statistics and the increasingly intertwined nature of spectral features at high Q's.

As a further remark, we would like to stress again the fact that results from Bayesian inference are always to be interpreted in a probabilistic nuance. For instance, we stated before that the oscillation mode Ω_1 lies in the interval (22.3; 25.2), with a probability of 95%. This interval, called *credibility interval*, is obtained by sorting the values of Ω_1, drawn from its posterior conditional to $k = 1$, and taking the two values below which we can find, respectively, the 2.5% and 97.5% of all simulated values of Ω_1. In practice, the values inside the interval are those with the highest density given the observed data and so the most credible. Classical confidence interval, obtained in the frequentist approach, does not have such a probabilistic interpretation. The interpretation of confidence intervals is that, if we imagine to repeat data sampling indefinitely under the same conditions and to build a confidence interval at

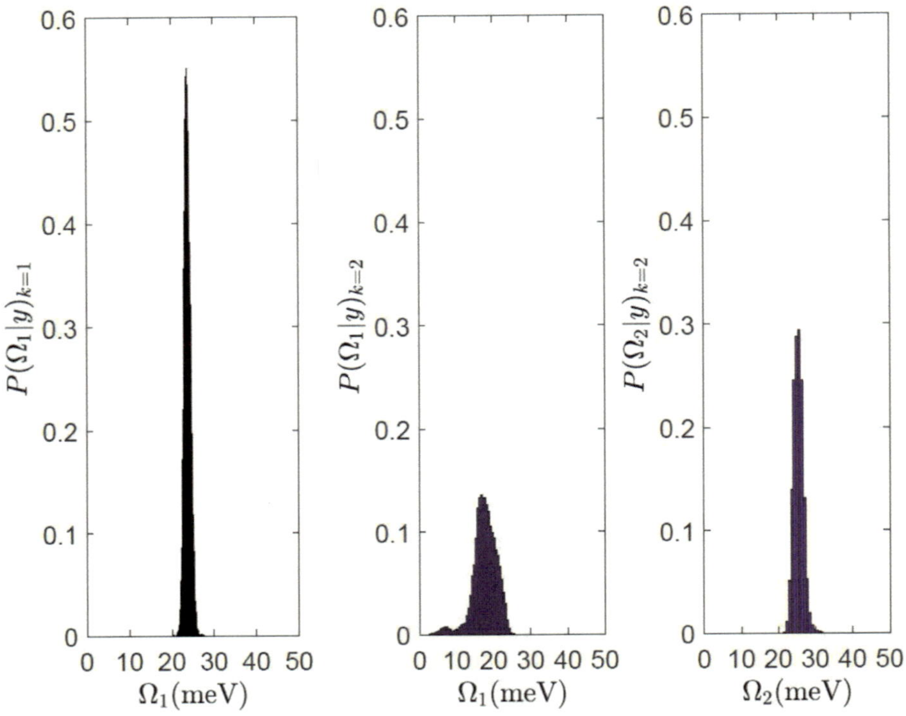

Figure 7.
Simulated posterior distributions for the excitation frequency Ω_1 and Ω_2 in the case of the model with $k = 1$ (panel on the left) and $k = 2$ (central and right panel) for liquid gold at a momentum transfer of $Q = 16\,\text{nm}^{-1}$.

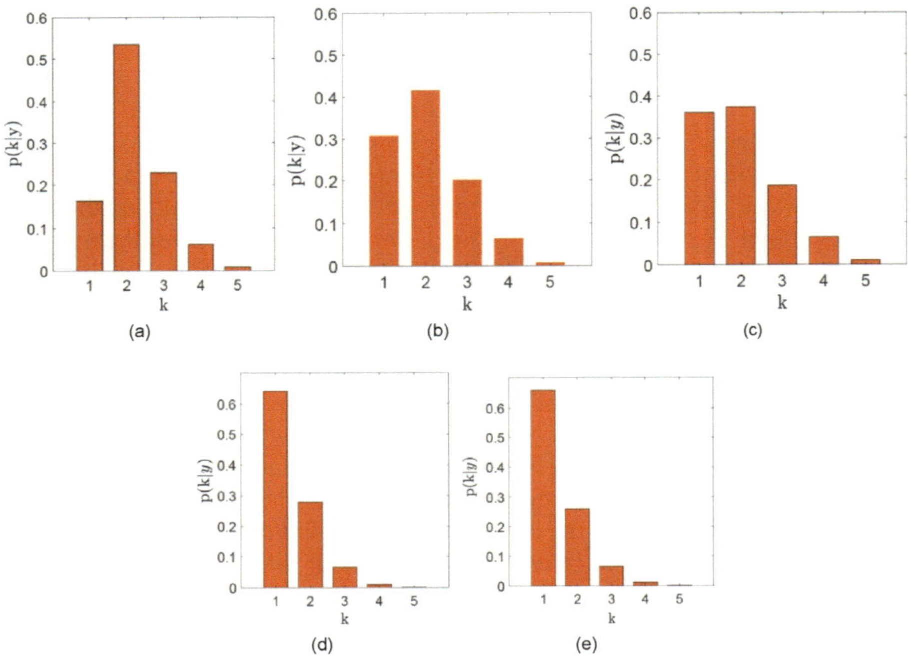

Figure 8.
Posterior probability for the number of k modes at different values of the momentum transfer Q in an inelastic scattering experiment performed on a gold nanoparticle suspension in water: (a) $Q = 3.5\,nm^{-1}$, (b) $Q = 5.5\,nm^{-1}$, (c) $Q = 7.5\,nm^{-1}$, (d) $Q = 9.5\,nm^{-1}$, and (e) $Q = 13.5\,nm^{-1}$. *Adapted with permission from Ref. [13], Copyright (2018) American Chemical Society.*

a certain $1-\alpha$ confidence level for each of the datasets, then (1–α)% of these confidence intervals will contain the true fixed value of the parameter. However, we have no guarantee that the single confidence interval, calculated on the basis of the only dataset actually observed, contains the true parameter value. We can only be confident [at the (1–α)% level] that it does so, since it comes from a set of intervals, (1–α)% of which do contain the parameter value. In practice, under a frequentist approach, data are random variables and give rise to random intervals that have a specific probability of containing the fixed, but unknown, value of the parameter. The single interval is also fixed and might or not contain the fixed parameter, but we cannot associate any probability measure to this possibility. In the Bayesian approach, the parameter is random in the sense that we have a prior belief about its value, while the interval can be thought of as fixed, once the data have been observed. In summary, the frequentist approach do provide a definition of confidence intervals, which, however, are endowed with a robust probabilistic ground only with respect to the hypothetic space of all possible repetitions of the measurement experiment but not with respect to the unique dataset at hand.

4.2 Bayesian inference in the time domain

Time correlation function decays can be modeled in terms of an expansion of the intermediate scattering function $I(Q,t)$ in exponentials, and the aim is often to determine the number of time decay channels that could be envisaged in the relaxation of $I(Q,t)$. In Ref. [15], the dynamics of polymer-coated gold nanoparticles in D_2O was tackled by neutron spin echo (NSE) scattering and analyzed within a Bayesian approach with the goal of establishing how many characteristic relaxations were present in a given spin echo time window and if they could be described by either simple or stretched exponentials or by a combination of the two. The data were assumed to be sampled by the following model:

$$y_i = \gamma \sum_{j=1}^{k} A_j \exp\left(-\left(\frac{t_i}{\tau_j}\right)^{\beta_j}\right) + \varepsilon_i, \qquad \text{for } i = 1, \dots, n \tag{21}$$

where γ is a proportionality constant possibly enabling a data normalization, k represents the number of exponential relaxations, A_j is the weight of the j-th component of the exponential mixture, τ_j its relaxation time, and β_j its stretching parameter. The ε_i, for $i = 1, \dots, n$, are random noises, accounting for statistical errors in the measurements. These are assumed to be independent and identically distributed with a normal distribution $\mathcal{N}(0, \upsilon\sigma_i^2)$, where σ_i is the measurement error corresponding to the i-th observation and υ is a proportionality constant. As a consequence, the likelihood of the data is a product of normal densities, each having mean $\gamma \Sigma_{j=1}^{k} A_j \exp\left(-\left(\frac{t_i}{\tau_j}\right)^{\beta_j}\right)$ and variance $\upsilon\sigma_i^2$.

The value of k is, obviously, unknown, and its determination is of great relevance. Therefore, also in this case, k is considered a stochastic variable to be estimated based on the data and conditional to all the other model parameters. Imagine that we have no clue about how many relaxations are necessary to describe the observed behavior of the time correlation function. However, we are aware that, in a case like this, the risk to over-parametrize the model is high, and we certainly know that, given the finite time window covered by the experiment and the limited number of experimental data, the number of relaxations should not be too large;

otherwise the results could be meaningless, hardly justifiable, and unlikely. Therefore, it seems a priori reasonable that k has a uniform distribution on the discrete values $k = 1, \cdots, k_{\max}$, where $k_{\max}$ is a small integer, as previously assumed when dealing with the number of excitations in the energy spectrum. Also, the relaxation times τ_j are supposed uniformly distributed on a continuous range of nonnegative values. The prior on A_j is tailored to ensure that the combination of relaxation terms fulfills the constraints $\Sigma_{j=1}^{k} A_j = 1$ and $A_j \geq 0$. The natural choice for the prior of $A = (A_1, A_2, \cdots, A_k)$ is, then, a Dirichlet density, which takes values on the standard simplex. A crucial prior is that of the stretching parameter β_j. This is specifically meant, in fact, to discern whether the relaxations in the given time window are simple exponential decays, stretched exponential decays, or a combination of the two. A simple exponential decay corresponds to $\beta_j = 1$, and thus a positive probability mass can be assigned to this specific value. The remaining probability can be assigned to β_j values within the interval $(0; 1)$. Therefore, a reasonable prior for β_j can be a mixed distribution made up of a probability mass in 1 and a continuous beta density, i.e., $\beta_j \sim \zeta \mathcal{B}(\kappa, \psi) + (1 - \zeta)\delta_{\beta_j,1}$, independently for $j = 1, \cdots, k$, where κ and ψ are parameters of the beta density, $\delta_{\beta_j,1}$ is an indicator function equal to 1 when $\beta_j = 1$ and 0 otherwise, and ζ is a weight denoting our prior support in favor of a stretched, rather than simple, exponential components. Once the $\zeta = 0$ and $\zeta = 1$ weights are, respectively, assigned to the sums of simple and stretched exponential terms in Eq. (21), other $0 < \zeta < 1$ weights will be associated to mixed combinations of these decay terms. In particular, a $\zeta = 0.5$ means that the j-th exponential can be either stretched or not with a priori the same probability, for all $j = 1, \cdots, k$. In addition, setting $\kappa = 1$ and $\psi = 1$ allows to assume that the stretching parameters are uniformly distributed on the interval $(0; 1)$. This corresponds to an uninformative prior giving a probability of 0.5 to both a stretched or unstretched component and, for a stretched component, assigning the same density to any value of β_j in $(0; 1)$. Obviously, more informative priors can be chosen, e.g., by assigning different values to κ, ψ and ζ, so to favor, for example, a Zimm or Rouse model specification (see discussion in Ref. [15]) when dealing with polymer dynamics. A similar prior probability can be adopted for the proportionality constant γ, i.e., mixed distribution made up of a continuous beta density and a probability mass in 1, corresponding to no need for a refinement of the data normalization process. Finally, the proportionality constant in the error variance, v, can be, for example, assumed to have a priori a gamma density so that only nonnegative values are allowed.

Let us consider one of the datasets in Ref. [15], representing the time correlation decay of a polymer solution of polyethylene glycol with a molecular weight of 2000D (PEG2000) as measured in a NSE scattering experiment and collected at a momentum transfer $Q = 0.091$ Å. Also in this case, we allowed for 10^5 sweeps of the algorithm and a burn-in of 10^4 sweeps, resulting in approximately 5/10 minutes of computing time. From the output of the MCMC-RJ algorithm, values for the discrete posterior distribution function of k are found in **Table 1**.

The most visited model is the one with two exponential functions. The fit is shown in the figure below (**Figure 9**).

The values reported in **Table 1** clearly show that the posterior distribution of k has a maximum. In fact, this is a general result (see, e.g., **Figures 8** and **10**).

When we model a spectroscopic dataset through a homogeneous mixture, e.g., a linear combination of exponentials, Lorentzians or DHO functions, the posterior distribution for the number of components always has at least a maximum, unless the data are so scarcely informative that the posterior for k simply reproduces the

k	$P(k\|y)\%$
1	8.47
2	61.83
3	23.91
4	4.41
5	1.12
6	0.26

Table 1.
Posterior distribution for the number of time correlation decay channels for a polymer solution of polyethylene glycol with a molecular weight of 2000D(PEG2000) as measured in a NSE experiment and collected at a momentum transfer $Q = 0.091\mathring{A}$.

prior, which might be uniform. In principle, when jumping in a more complicated model characterized by a larger number of parameters, the χ^2 tends to decrease, and the likelihood tends to increase. However, according to the Bayes theorem, the posterior for k is computed averaging the likelihood over all the parameters value (see Eq. (12)). Therefore, models that are under-parametrized will perform poorly on average since they just cannot fit the data well enough and have a small likelihood, while models that are over-parametrized will also perform poorly on average, because the subset of the parameter space that fit the data well (and where the likelihood is high) becomes tiny compared to the whole volume of the parameter space. This means that adding components to the mixture model increases the posterior distribution of k only until the increment in the likelihood more than compensates for the augmentation of the "wasted" parameter space; overall the competition of these effects ensures the presence of a maximum in $P(k|y)$. It is worth noticing that assuming a model with more free parameters does not necessarily mean a better fit, once the likelihood has saturated. To see this, we report here below (**Figure 11**) the fit we get with a number of relaxation channels $k \neq 2$. We can observe how the fit with three relaxation components or more is not better than the one more supported by the available data and estimated by the MCMC-RJ algorithm. Moreover it is *insane* and hopeless to confer a distinct physical meaning to each one of the corresponding characteristic relaxation times.

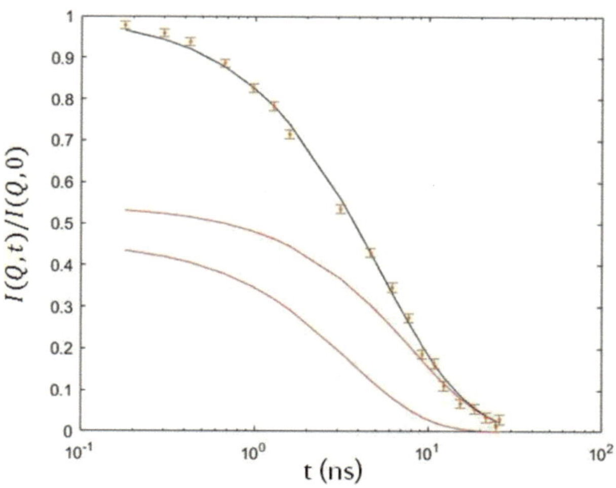

Figure 9.
$I(Q,t)/I(Q,0)$ vs. time (ns). The black line is the best fit as determined with the RJ-MCMC. The two red lines are the two exponential components.

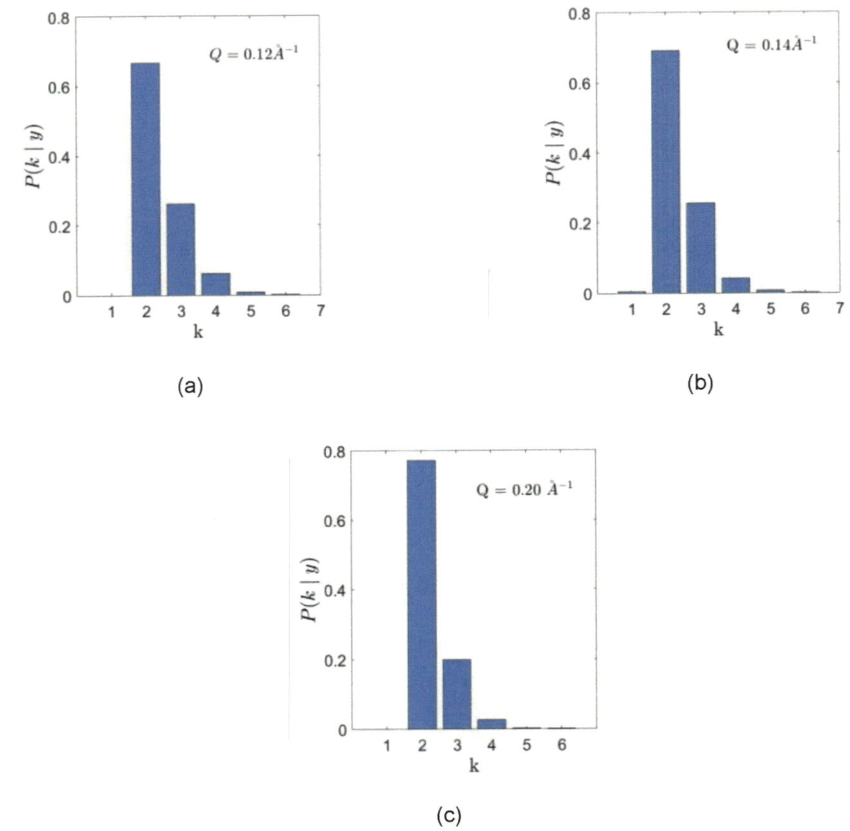

Figure 10.
Posterior probability for the number of k modes at different values of the momentum transfer Q in an NSE experiment performed on polymer solution of polyethylene glycol with a molecular weight of 2000D (PEG2000) *in* D_2O.

Let us introduce a quantity which could resemble the χ^2, namely:

$$s^2 = \sum_{i=1}^{n} \frac{\left(y_i - y_{calc}\right)^2}{\sigma_i^2}, \tag{22}$$

which measures the distance between the experimental data and the best fit determined with the RJ-MCMC algorithm, where n is the number of experimental observations, y_i are the experimental data, y_{calc} are the best fit calculated values, σ_i are the experimental errors. This variable differs from the usual χ^2 as the model parameters are not estimated by least squares minimization, but are the averages, of the corresponding marginal posterior distributions. Nevertheless we can use this quantity to show what follows. If we calculate the quantity in Eq. (22) for each value of k, we get for s^2 the values reported in **Table 2** which indicates an overall decrease upon increasing the number of exponentials. Actually, s^2 does not strictly decrease with the numbers of parameters, because, as mentioned before, the fit is not calculated with parameter values which minimize the χ^2. If, for example, in particular situations (e.g., for $k = 3$), the algorithm faces some challenges in determining a parameter and its posterior distribution is *very broad* and slowly decaying, the average of this parameter could be severely affected by the presence of these sizable distribution tails. In these cases, the mode of the distribution should be used instead to estimate the parameter.

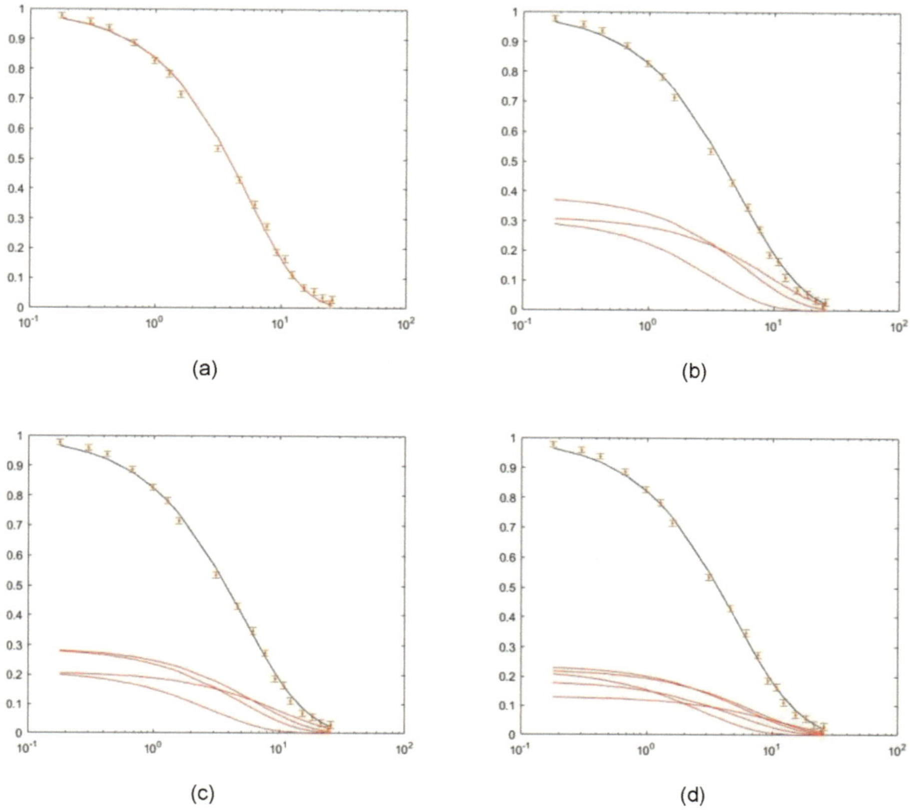

Figure 11.
$I(Q,t)/I(Q,0)$ vs. line (ns). The black line is the best fit as determined with the RJ-MCMC. The two red lines are the two exponential components. (a) $k = 1$, (b) $k = 3$, (c) $k = 4$, and (d) $k = 6$.

k	s^2
1	13.26
2	9.07
3	10.92
4	9.00
5	8.57
6	8.57

Table 2.
Values of the quantity s^2 as defined in Eq. (22) calculated for the different values of k and considering the averages of the model parameter posterior distribution.

Nevertheless, s^2 shows that even with a distance between experimental and fitted values which is effectively decreasing as the number of parameters increases, the most probable model from **Table 1** is the one with $k = 2$ and not the one with $k = k_{\max}$. The effect of the razor is evident. It can also be noted that the fit with $k = 2$ is not only the most probable but it is also the best (in the sense that it is much better than the one with $k = 1$ and it is not worse than those obtained using a larger number of exponentials). More interestingly, we observe also that increasing the number of parameters, the χ^2 (or any other measure of the distance between the fitted and the observed data) is not decreasing much for large values of k, because obviously at the end, this quantity is going to saturate (and so does the likelihood). The fit with $k = 2$ determines a value of s^2, which is not too different from the one we get with $k = 6$. Incidentally, as it is largely discussed in Ref. [15], the model with two relaxation channels has also a perfectly plausible and consistent explanation, which would not be possible if a more complicated model were chosen.

In summary, we have here shown some of the opportunity offered by a Bayesian inference analysis of experimental results and, in particular, those obtained with spectroscopic methods. As possible future development, it appears very promising the opportunity of applying similar methods to the joint analysis of complementary time or frequency-resolved measurements. Also, we can envisage the use of more informative priors implementing the fulfillment of sum rules of the spectra or any other known physical constraint of the measurement. We are confident that, in the long run, these methods will improve the rigor of routine data analysis protocols, supporting a probability-based, unprejudiced interpretation of the experimental outcome.

This work used resources of the National Synchrotron Light Source II, a US Department of Energy (DOE) Office of Science User Facility operated for the DOE Office of Science by Brookhaven National Laboratory under Contract No. DE-SC 0012704. The open access fee was covered by FILL2030, a European Union project within the European Commission's Horizon 2020 Research and Innovation programme under grant agreement N°731096.

Acknowledgements

We would like to thank U. Bafile, E. Guarini, F. Formisano, and M. Maccarini for the very stimulating discussions.

Author details

Alessio De Francesco[1,2*], Alessandro Cunsolo[3] and Luisa Scaccia[4]

1 CNR—IOM, Grenoble, France

2 Institut Laue-Langevin, Grenoble, France

3 National Synchrotron Light Source-II, Brookhaven National Laboratory, Upton, NY, USA

4 Università di Macerata, Macerata, Italy

*Address all correspondence to: defrance@ill.fr

References

[1] Boon JP, Yip S. Molecular Hydrodynamics. Mineola, NY: Dover Publication Inc.; 1980

[2] Hansen J-P, McDonald IR. Theory of Simple Liquids. New York: Academic Press; 1976

[3] Balucani U, Zoppi M. Dynamics of the Liquid State. Vol. 10. Oxford: Clarendon Press; 1995

[4] Copley J, Lovesey S. The dynamic properties of monatomic liquids. Reports on Progress in Physics. 1975;**38**:461

[5] Scopigno T, Ruocco G, Sette F. Microscopic dynamics in liquid metals: The experimental point of view. Reviews of Modern Physics. 2005;**77**:881

[6] Berne BJ, Pecora R. Dynamic Light Scattering: With Applications to Chemistry, Biology, and Physics. New York: Dover Publications, Inc.; 2000

[7] Fleury PA, Boon JP. Brillouin scattering in simple liquids—argon and neon. Physical Review. 1969;**186**:244

[8] Cunsolo A, Pratesi G, Verbeni R, Colognesi D, Masciovecchio C, Monaco G, et al. Microscopic relaxation in supercritical and liquid neon. The Journal of Chemical Physics. 2001;**114**: 2259

[9] Gilks WR, Richardson S, Spiegelhalter DJ. Markov Chain Monte Carlo in Practice. London: Chapman & Hall/CRC; 1996

[10] Tierney L. Markov chains for exploring posterior distributions. The Annals of Statistics. 1994;**22**:1701

[11] Green PJ. Reversible jump Markov chain Monte Carlo computation and Bayesian model determination. Biometrika. 1995;**82**:711

[12] De Francesco A, Guarini E, Bafile U, Formisano F, Scaccia L. Bayesian approach to the analysis of neutron Brillouin scattering data on liquid metals. Physical Review E. 2016;**94**: 023305

[13] De Francesco A, Scaccia L, Maccarini M, Formisano F, Zhang Y, Gang O, et al. Damping off terahertz sound modes of a liquid upon immersion of nanoparticles. ACS Nano. 2018;**12**:8867

[14] De Francesco A, Scaccia L, Formisano F, Maccarini M, De Luca F, Parmentier A, et al. Shaping the terahertz sound propagation in water under highly directional confinement. Physical Review B. 2020; **101**:05436

[15] De Francesco A, Scaccia L, Lennox RB, Guarini E, Bafile U, Falus P, et al. Model-free description of polymer-coated gold nanoparticle dynamics in aqueous solutions obtained by Bayesian analysis of neutron spin echo data. Physical Review E. 2019;**99**: 052504

[16] Parmentier A, Maccarini M, De Francesco A, Scaccia L, Rogati G, Czakkel O, et al. Neutron spin echo monitoring of segmental-like diffusion of water confined in the cores of carbon nanotubes. Physical Chemistry Chemical Physics. 2019;**21**:21456

[17] Bemardo JM. Philosophy of statistics. In: Bandyopadhyay PS, Forster MR, editors. Handbook of the Philosophy of Science. Vol. 7. Amsterdam: North-Holland; 2011. pp. 263-306

[18] Berger JO, Jefferys WHA. The application of robust Bayesian analysis to hypothesis testing and Occam's razor. Journal of the Royal Statistical Society, Series A. 1992;**1**:17

[19] Jefferys WH, Berger JO. Ockham's Razor and Bayesian Analysis. American Scientist. 1992;**80**:64

[20] MacKay D. Information Theory, Inference and Learning Algorithms. Cambridge: Cambridge University Press; 2003

[21] Chib S, Greenberg E. Understanding the metropolis-hastings algorithm. The American Statistician. 1995;**49**:327

[22] Roberts GO, Gelman A, Gilks WR. Weak convergence and optimal scaling of random walk metropolis algorithms. The Annals of Applied Probability. 1997; **7**:110

[23] Cowles MK, Carlin BP. Markov chain Monte Carlo convergence diagnostics: A comparative review. Journal of the American Statistical Association. 1996;**91**:883

[24] Guarini E, Bafile U, Barocchi F, De Francesco A, Farhi E, Formisano F, et al. Dynamics of liquid Au from neutron Brillouin scattering and ab initio simulations: Analogies in the behavior of metallic and insulating liquids. Physics Review. 2013;**B88**: 104201

[25] De Francesco A, Scaccia L, Maccarini M, Formisano F, Guarini E, Bafile U, et al. Interpreting the terahertz spectrum of complex materials: The unique contribution of the Bayesian analysis. Materials. 2019;**12**:2914

Chapter 3

Atomic Dynamics in Real Space and Time

Takeshi Egami

Abstract

Atomic and molecular dynamics in strongly disordered matter, such as liquid, cannot be fully described in terms of phonons, because they are marginalized and often overdamped. Their dynamic and transport properties depend on local atomic rearrangements which are strongly correlated. To describe such local dynamics, the usual representation in momentum (Q) and energy (E) space in terms of the dynamic structure factor, $S(Q, E)$, is not effective. We discuss an alternative approach in real space (r) and time (t), with the van Hove function, $G(r, t)$, and show how this approach facilitates understanding of real-space local dynamics of liquids and other disordered systems in the length scale of Å and time scale of pico-second.

Keywords: local atomic dynamics, van Hove function, liquid, glass, correlated dynamics

1. Introduction

In crystalline solids, phonons are the elementary excitations of lattice dynamics. They can be observed with well-defined dispersions in the dynamic structure factor, $S(\mathbf{Q}, E)$, where $\mathbf{Q}$ is the momentum exchange and E (= $\hbar\omega$) is the energy exchange in scattering. The $S(\mathbf{Q}, E)$ can be measured by inelastic X-ray scattering (IXS) as discussed in this book or by inelastic neutron scattering (INS). However, in strongly disordered materials, shortwave phonons have very short mean free path and lifetime. Only long-wave phonons, for which the atomic structure is irrelevant and the material acts as a continuum elastic body, propagate over some distance. However, the total spectral weight of long-wave phonons is small, because the phonon density of states increases as E^2. For this reason, a majority of phonons are overdamped, and some of them are localized, particularly in liquid.

Atomic dynamics in liquid and soft matter is usually studied by quasi-elastic scattering (QES) at low Q, where the QES intensity is dominated by the self-correlation and the auto-correlation of the same atom. The energy width of QES is proportional to DQ^2, where D is the diffusion constant. For hydrogen, which has a very large incoherent cross section, the value of D determined by QES at low Q is the isotopic diffusivity, D_i, determined by the diffusion profile of isotopes measured by sectioning. The chemical diffusivity, D_c, which characterizes the flow of elements responding to concentration gradient, is usually different from D_i, because of the backflow of diffusing atoms. The ratio, $f = D_c/D_i$, is known as the correlation factor and describes how diffusive atomic jumps are correlated.

This example illustrates the importance of studying the distinct atomic correlation and the dynamic correlation among different atoms. The importance of knowing the distinct terms of the correlation function is beginning to be recognized [1, 2], but the difficulty of measurement delayed advances. Only recently the progress in IXS instrumentation [3, 4] and the advent of pulsed neutron sources [5] made it feasible to measure $S(Q, E)$ over wide ranges of Q and E in a reasonable time and opened the possibility of garnering information on distinct atomic correlations.

Conversely, even though the substantial progress in instrumentation allowed us to collect a much larger amount of data which contain vastly richer information, the IXS data are mostly processed in the same way, just by focusing on the phonon dispersion and its width [6]. Similarly, usually only the diffusivity is obtained from the QES data. What is missing is the analysis of the *diffuse* IXS intensity which is usually discarded as background. This situation is a perfect analog of the structural analysis of disordered crystals. The conventional methods of crystallographic analysis focus only on the Bragg peaks, and the diffuse scattering intensity is processed separately to determine the short-range order. However, by combining the Bragg and diffuse intensities as the total scattering intensity and Fourier-transforming it into the atomic pair distribution function (PDF), precise local structural information can be obtained [7]. In the same manner, by including the diffuse IXS intensity, valuable information on local dynamics can directly be obtained.

2. Van Hove function

2.1 Definition

After the correction for absorption and normalization, the IXS intensity, $I(Q, E)$, is reduced to $S(Q, E)$ [8],

$$S(\mathbf{Q}, E) = \frac{1}{2\pi\hbar N}\sum_{i,j}\int \left\langle \exp\left[\mathbf{Q}\cdot\left(\mathbf{r_i}(0) - \mathbf{r_j}(t)\right)\right]\right\rangle \exp\left(-\frac{E}{\hbar}t\right)dt, \tag{1}$$

where $\boldsymbol{r_i}(t)$ is the position of the atom i at time t. It is useful to separate it into the self-part ($i = j$), $S_s(\mathbf{Q}, E)$, and the distinct-part ($i \neq j$), $S_d(\mathbf{Q}, E)$.

$$S(\mathbf{Q}, E) = S_s(\mathbf{Q}, E) + S_d(\mathbf{Q}, E). \tag{2}$$

For isotropic matter such as liquid, we use the spherical average,

$$S(Q, E) = \frac{1}{4\pi}\int S(\mathbf{Q}, E)d\Omega, \tag{3}$$

where $Q = |\mathbf{Q}|$ and Ω is the solid angle in Q space. Upon the Fourier transformation into the time-domain, we obtain the intermediate scattering function [8],

$$F(Q, t) = \int S(Q, E)\exp\left(i\frac{E}{\hbar}t\right)dE, \tag{4}$$

which has been widely used in the analysis of soft matter dynamics [9–11]. Another step of the Fourier transformation, this time from momentum space to real space, leads to the van Hove function [12],

$$G(r,t) = 1 + \frac{1}{2\pi^2\rho_0 r}\int F(Q,t)\sin(Qr)QdQ$$
$$= \frac{1}{4\pi\rho_0 N r^2}\sum_{i,j}\delta\left(r - \left|\mathbf{r}_i(0) - \mathbf{r}_j(t)\right|\right). \tag{5}$$

Again it is useful to divide it into the self-part, $G_s(r,t)$, and the distinct-part, $G_d(r,t)$. Although the van Hove function has been known for a long time, its experimental determination has rarely been done [13] because it requires $S(Q,E)$ to be known over wide ranges of Q and E. Only recently it became practical [14, 15] because of the progress in instrumentation as noted above and discussed below.

In the regular X-ray diffraction measurement, the energy resolution is of the order of 1 eV, far greater than the phonon excitation energies. Therefore what is measured is the energy-integrated intensity,

$$S(Q) = \int S(Q,E)dE, \tag{6}$$

which leaves only the same-time ($t = 0$) contribution in Eq. (1). Therefore its Fourier transform, the PDF,

$$g(r) = 1 + \frac{1}{2\pi^2\rho_0 r}\int [S(Q) - 1]\sin(Qr)QdQ, \tag{7}$$

is equal to $G(r, 0)$. In other words the van Hove function describes how the PDF, the snapshot correlation function, decays with time.

2.2 Evolution with time

At a short time scale ($\sim$0.1 ps), atomic motions are ballistic, but after atoms leave the neighbor cage, they become diffusive. Then the self-term of the van Hove function in the diffusive regime should be

$$G_s(r,t) = \frac{1}{\rho_0}\left(\frac{1}{4\pi D_i t}\right)^{\frac{3}{2}}\exp\left(-\frac{r^2}{4D_i t}\right). \tag{8}$$

Therefore the self-diffusion coefficient, D_i, can be determined from the self-part of the van Hove function. The double Fourier transform of Eq. (8) is

$$S_s(Q,\omega) = \frac{D_i Q^2/\pi}{\omega^2 + \left(D_i Q^2\right)^2}, \tag{9}$$

which can be measured by QES and is routinely used for determining D_i. In the case of quasi-elastic neutron scattering (QENS) from hydrogen, because the incoherent cross section of hydrogen is so large, the scattering intensity is totally dominated by $S_s(Q,\omega)$, and the measurement readily yields the value of D_i. In general, however, there can be some contributions from the distinct-part to the low-angle QES, which can make the measurement inaccurate. On the other hand in the van Hove function the self-part is cleanly separated, at least initially, resulting in more accurate determination of D_i.

The decay of the first peak of the PDF with time describes how the nearest neighbor shell of an atom, known as the first-neighbor cage, disintegrates with time. The van Hove function depicts this decay nicely and can relate the time scale of decay to the topological relaxation time and to viscosity as shown below.

The early prediction on the distinct-part of the van Hove function was that it could be expressed by the convolution of the PDF by the self-part (Eq. (8)) [16]. Then the QES width should be equal to DQ^2, by Eq. (9). But de Gennes noted that QES becomes anomalously narrow in the vicinity of the first peak in $S(Q)$ [17]. He suggested that this phenomenon, now known as the de Gennes narrowing, was due to the collective nature of the dynamics represented by the first peak in $S(Q)$. Since then it became customary to equate the observation of the de Gennes narrowing to the confirmation of collective excitations. A recent study [18], however, showed otherwise. It was found that even in high-temperature liquid, in which atomic motions are uncorrelated, the decay time of $G(r, t)$ depends linearly on distance. In general,

$$\tau(r) = \tau_0 + \tau_r \left(\frac{r}{r_1}\right)^{\chi}, \tag{10}$$

where r_1 is the position of the nearest neighbor peak of the PDF. As shown in **Figure 1**, the exponent χ depends on dimensionality d approximately as $\chi = (d\text{-}1)/2$; thus $\chi = 1$ for three dimensions. This is because at large r, each PDF peak describes not just one atomic distance but many. Therefore its decay with time does not correspond to the single atom dynamics. The number of pairs of atoms in each peak, N_r, is proportional to the surface area, $4\pi r^{d-1}$. Then its fluctuation is proportional to $\sqrt{N_r} \sim r^{(d-1)/2}$; therefore $\chi = (d\text{-}1)/2$. Now the first peak of $S(Q)$ represents the medium-range part of the PDF, beyond the first peak [19], so its decay is slow, reflecting the behavior of the PDF beyond the first peak. This argument proves that the de Gennes narrowing does not necessary imply collective excitations but can be just the natural consequence of geometry.

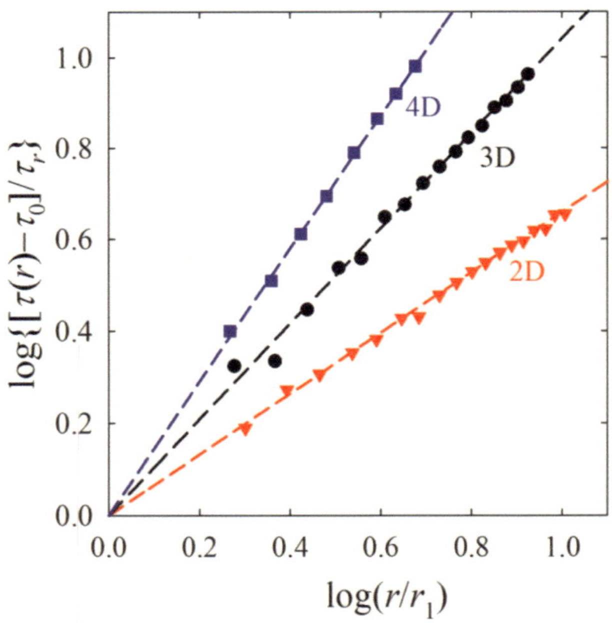

Figure 1.
The r-dependent relaxation time $\tau(r)$ for model liquid iron in 2D (red triangle) $\chi = 0.66$, 3D (black circle) $\chi = 1.04$, and 4D $\chi = 1.45$, beyond the first peak. The data points are shown in the form of $\log\{[\tau(r)\text{-}\tau_0]/\tau_r\}$ versus $\log(r/r_1)$ to highlight χ from the expected power law dependence. The short dashed lines serve as guides to the eye [18].

3. Local dynamics of water and aqueous solution of salt

3.1 Van Hove function of water

Figure 2 shows the $S(Q, E)$ of water at room temperature, determined by the IXS experiment at the beam line XL35 of the SPring-8 facility [14]. Earlier IXS experiments to observe phonons did not cover the Q space much beyond 1 Å^{-1} [6, 20, 21]. The $S(Q, E)$ is dominated by QES, and as is given it is not easy to garner useful information without extensive modeling. Converting the data into the van Hove function makes local dynamics directly visible as shown in **Figure 3**. Because hydrogen is almost invisible to X-rays, the van Hove function is dominated by oxygen–oxygen correlation. To minimize the termination error for stopping the integration by Eq. (4) at a maximum Q value, $F(Q, t)$ can be extended to large Q

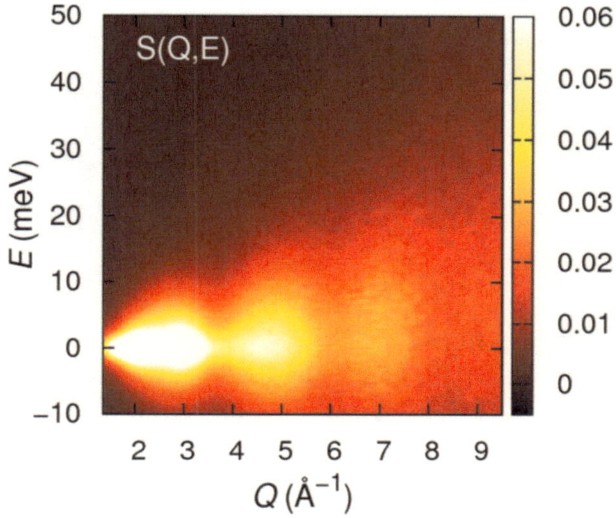

Figure 2.
The S(Q, E) of water at room temperature, determined by the IXS experiment at the beam line XL35 of the SPring-8 facility [14].

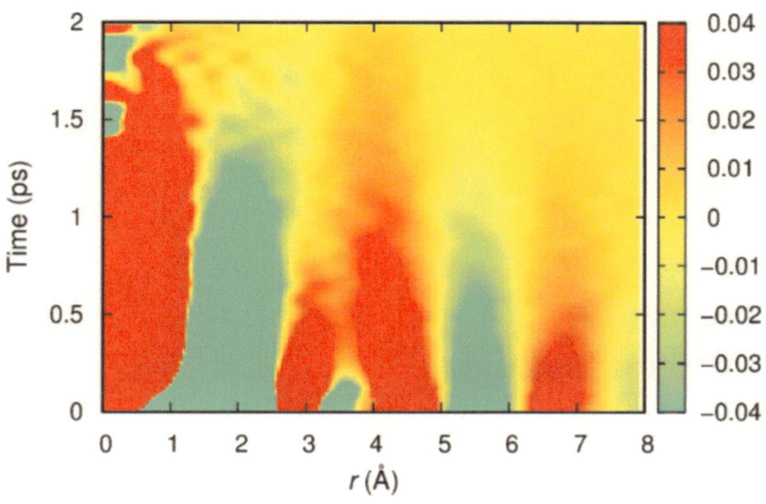

Figure 3.
The van Hove function of water [14].

by adding $S(Q)$ exp.$(-D(Q)Q^2t)$, which is justified for the self-correlation function [14].

In **Figure 3** the data at $t = 0$ is the snapshot PDF which can be obtained by the conventional diffraction measurement. At $t = \infty$ $G(r, \infty) = 1$, so that $G(r, t) - 1$ describes the correlation. The decay of the PDF to $G(r, \infty) = 1$ is not uniform, with each peak behaving in different ways. In particular the first peak moves away, while the second peak moves in, indicating that the local dynamics is highly correlated. As the nearest neighbor moves away, the second neighbor comes in to take its place to maintain the coordination unchanged. The area of the first peak above $G(r, t) = 1$ shows a two-step decay,

$$A(t) = A_1 e^{-(t/\tau_1)^{\gamma_1}} + A_2 e^{-(t/\tau_2)^{\gamma_2}}. \tag{11}$$

The first term ($\tau_1 = 0.32$ ps) describes the ballistic motion of the atom, whereas the second term with the temperature-dependent τ_2 describes the change in molecular bond. Earlier through molecular dynamics (MD) simulations, it was found that the time scale of losing one nearest neighbor, τ_{LC}, is directly related to viscosity through $\tau_{LC} = \tau_M = \eta/G_\infty$, where τ_M is the Maxwell relaxation time, η is viscosity, and G_∞ is instantaneous shear modulus [22]. By relating τ_2 to τ_{LC} through simulation (for water $\tau_2 = \tau_{LC}$), this relationship was proven for water [14, 23].

3.2 Self-diffusion

The portion of the van Hove function near $r = 0$ describes the self-correlation, $G_s(r, t)$. Indeed it follows Eq. (8) quite well for water as shown in **Figure 4** [24].

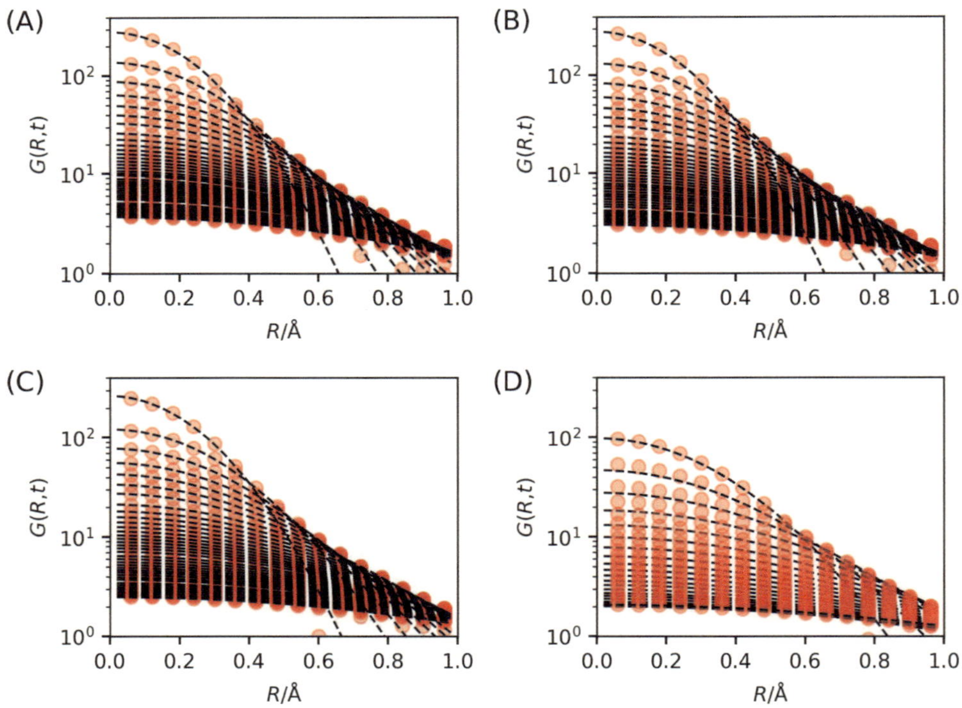

Figure 4.
The self-part of the van Hove function for water at (A) 285 K, (B) 295 K, (C) 310 K, and (D) 318 K. (circles) experimental data and (dashed line) the result of fitting by Eq. (8).

However, the values of diffusivity determined from Eq. (8) vary from the values obtained by other methods [24]. The origin of this discrepancy is yet to be determined.

3.3 Van Hove function of salty water

About 70% of the earth is covered by salty water, and 80% our body is also made of salty water. Therefore it is important to know how salt affects the properties of

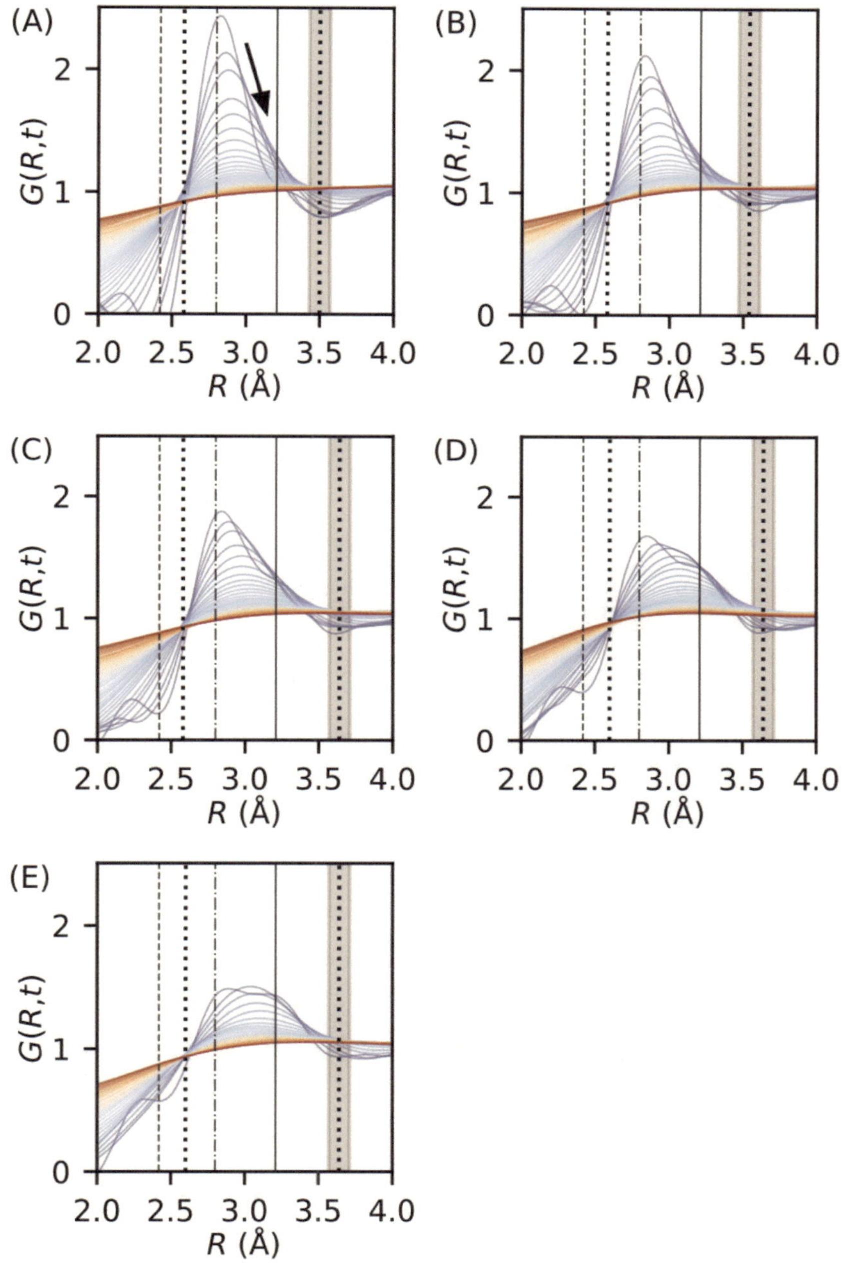

Figure 5.
The van Hove functions around the first-neighbor correlation peak, $R \sim 2.9$ Å: (A) pure water, (B) m = 0.75 mol/kg, (C) 1.5 mol/kg, (D) 2.26 mol/kg, and (E) 3.0 mol/kg. The solid lines at R = 3.21 Å show the R_{O2-} + R_{Cl-}. The dashed line at R = 2.42 Å shows the R_{O2-} + R_{Na+}. The dash-dotted line at R = 2.8 Å shows the R_{O2-} + R_{O2-}. The range between the dotted lines ($\mathbf{R_1}'$ and $\mathbf{R_1}''$) was used to calculate the area, A(t), of the first neighbor. The upper limit of this range is changed within the gray-shaded area to estimate the uncertainties [25].

water, such as viscosity. We studied the local dynamics of aqueous solution of NaCl up to 2 mol/kg by IXS [24] using the BLX-43 beam line of SPring-8 which has as many as 24 analyzer crystals. With this setup a dataset similar to that shown in **Figure 2** can be collected in 12 h.

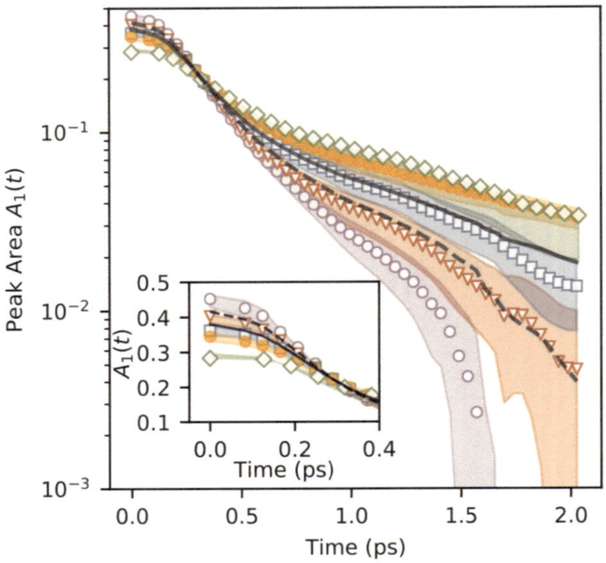

Figure 6.
Temporal evolution in the area of the first-neighbor peak, $A_1(t)$, and the enlarged view (inset): (open circles) pure water, (triangles) m = 0.75 mol/kg, (squares) 1.5 Mol/kg, (closed circles) 2.26 mol/kg, and (diamonds) 3.0 mol/kg. The shaded areas represent uncertainties of each dataset. The solid and dashed lines represent the linear combination of time evolution for m = 0 and 2.26 mol/kg [25].

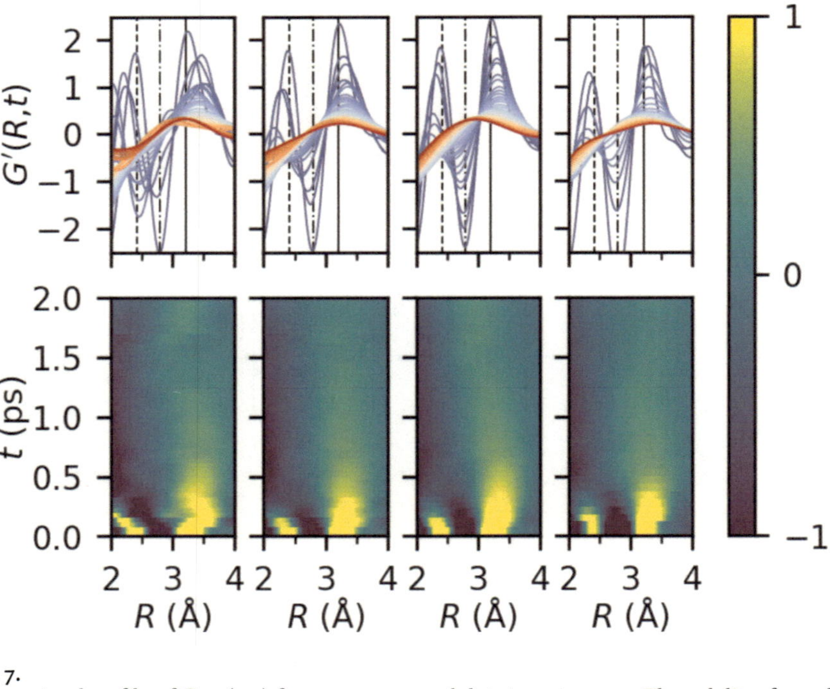

Figure 7.
One-dimensional profiles of $\mathbf{G}_{w-s}(\mathbf{r},\mathbf{t})$ for $0 < t < 2$ ps and their intensity maps. The molality of sample is 0.75, 1.5, 2.26, and 3.0 mol/kg from the left to the right. The solid lines, dashed lines, and the dash-dotted lines in the top figures represent R = 3.21 Å ($R_{O2-} + R_{Cl-}$), R = 2.42 Å ($R_{O2-} + R_{Na+}$), and R = 2.8 Å ($R_{O2-} + R_{O2-}$), respectively.

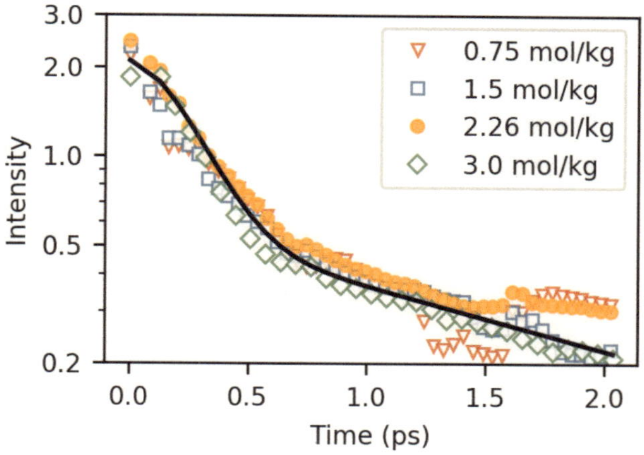

Figure 8.
Time evolution of peak height at around R = 3.21 Å. the solid line shows the result of fitting using two (compressed) exponential functions (Eq. (12)) [25].

As shown in **Figure 5**, the height of the first peak of the van Hove function is reduced by salt. The time dependence of the area of the first peak above $G(r, t) = 1$, shown in **Figure 6**, demonstrates that the addition of salt increases the slow decaying component. Furthermore it is possible to decompose the van Hove function to that of the water–water correlation, $G_{w\text{-}w}$, and that of the water-salt correlation, $G_{w\text{-}s}$,

$$G(r,t) = w_{w-w}G_{w-w}(r,t) + w_{w-s}G_{w-s}(r,t), \tag{12}$$

where $w_{w\text{-}w}$ and $w_{w\text{-}s}$ are the X-ray scattering weight for each component. The salt-salt correlation was neglected because the concentration of salt was low. If we assume that $G_{w\text{-}w}$ is the same as for pure water, we can determine $G_{w\text{-}s}$ from Eq. (12). As shown in **Figure 7**, $G_{w\text{-}s}$ is almost the same for all concentrations. The decay of the area of the sub-peak at 3.2 Å, corresponding to the Cl-O distance, is also the same for all concentrations as shown in **Figure 8**, proving the effect of salt on dynamics is local.

4. Limitations of the method

For the determination of the van Hove function, the current setup of IXS is ideally suited to the study of local dynamics in the time scale of 0.1–2 ps and length scale up to 5 Å. The energy resolution ($\sim$ 1.5 meV) sets the long-time limit to 2 ps. The effect of resolution is mitigated by the data analysis, by correcting the intermediate function for resolution,

$$F(Q,t) = F_{obs}(Q,t)/F_{res}(Q,t), \tag{13}$$

where $F_{res}(Q, t)$ is the Fourier transformation (Eq. (3)) of the energy resolution function. However, when $F_{res}(Q, t)$ becomes too small at long t, this correction is no longer sufficient. This represents a severe limitation for the IXS-derived van Hove function. To go beyond this limit, we have to resort either to neutron scattering which offers better energy resolution or to develop the method of X-ray photon correlation spectroscopy (XPCS) with free-electron X-ray laser [26]. At the moment, because the method was only recently proven to be feasible, there are many low-hanging fruits which we are busy collecting.

5. Concluding remarks

As the instrumentation for IXS is improved, it became possible to carry out many *Q-E* scans in a relatively short time, covering wide *Q-E* space. This enables us to convert the dataset of the dynamic structure factor $S(Q, E)$ into the van Hove function $G(r, t)$ which describes dynamics in real space and time. This new capability allows us to visualize local dynamics directly rather than through the modeling for $S(Q, E)$. Even though the van Hove function is just the Fourier transform, in order to determine it, we have to measure $S(Q, E)$ over a wide *Q-E* space, which forces us to collect much more information than we normally do. This alone brings us to a new territory. Furthermore, by visualizing it, we gain much intuition. For instance, the behavior of the first and second peaks of water shown in **Figure 3** was totally unexpected. Only after seeing it, we understand right away what this means, but no one would have anticipated it before it is seen. It is fully expected that such surprises will occur when this approach is applied further to the study of local dynamics in liquids, glasses, and soft matter in general.

Acknowledgements

The results reviewed in this article were obtained through the extraordinary effort and dedication by the team members of the author, W. Dmowski, T. Iwashita, and Y. Shinohara, in particular. He is also grateful to A. Q. R. Baron for his invaluable help during the experiment at SPring-8. This work was supported by the Department of Energy, Office of Science, Basic Energy Sciences, Materials and Science and Engineering Division.

Conflict of interest

The author declares no conflict of interest, financial or otherwise.

DOI: http://dx.doi.org/10.5772/intechopen.88334

Author details

Takeshi Egami[1,2]

1 University of Tennessee, Knoxville, TN, USA

2 Oak Ridge National Laboratory, Oak Ridge, TN, USA

*Address all correspondence to: egami@utk.edu

References

[1] Arbe A, Colmenero J, Farago B, Monkenbusch M, Buchenau U, Richter D. Intermediate length scale dynamics in glass forming polymers: Coherent and incoherent quasielastic neutron scattering results on polyisobutylene. Chemical Physics. 2003;**292**:295-309

[2] Novikov VN, Schweizer KS, Sokolov AP. Coherent neutron scattering and collective dynamics on mesoscale. The Journal of Chemical Physics. 2013;**138**: 164508

[3] Baron AQR. High-resolution inelastic X-ray scattering I: Context, spectrometers, samples, and superconductors. In: Synchrotron Light Sources and Free-Electron Laser. Switzerland: Springer International Publishing; 2015

[4] Baron AQR, Tanaka Y, Goto S, Takeshita K, Matsushita T, Ishikawa T. Journal of Physics and Chemistry of Solids. 2000;**61**:461-465

[5] Mason TE, Abernathy D, Anderson I, Ankner J, Egami T, Ehlers G, et al. The spallation neutron source in oak ridge: A powerful tool for materials research. Physica B. 2006;**385**:955-960

[6] Scopigno T, Ruocco G. Microscopic dynamics in liquid metals: The experimental point of view. Reviews of Modern Physics. 2005;**77**:881-933

[7] Egami T, Billinge SJL. Underneath the Bragg Peaks: Structural Analysis of Complex Materials; 2nd ed., Pergamon Materials Series Vol. 16. Oxford, UK: Pergamon Press, Elsevier Ltd.; 2012

[8] Lovesey SW. Theory of Neutron Scattering from Condensed Matter. Oxford, UK: Oxford University Press; 1984

[9] Mezei F, Knaak W, Farago B. Neutron spin-echo study of dynamic correlations near liquid-glass transition. Physica Scripta. 1987;**T19B**: 363-368

[10] Richter D, Monkenbusch M, Arbe A, Colmenero J. Neutron spin echo in polymer systems. Advances in Polymer Science. 2005;**174**:1-221

[11] Ruta B, Baldi G, Chuskin Y, Ruffle B, Cristoflini L, Fonatan A, et al. Revealing the fast atomic motion of network glasses. Nature Communications. 2014;**5**:3939

[12] Van Hove L. Correlation in space and time and born approximation scattering in systems of interacting particles. Physics Review. 1954;**95**:249-262

[13] Dahlborg U, Gudowski W, Davidovic M. Van Hove correlation functions from coherent neutron inelastic scattering. Journal of Physics: Condensed Matter. 1989;**1**:6173-6179

[14] Iwashita T, Wu B, Chen W, Tsutsui S, Baron AQR, Egami T. Seeing real-space dynamics of liquid water through inelastic X-ray scattering. Science Advances. 2017;**3**:e1603079

[15] Kawakita Y, Kikuchi T, Inamura Y, Tahara S, Maruyama K, Hanashima T, et al. Anomaly of structural relaxation in complex liquid metal of bismuth—Dynamic correlation function of coherent quasi-elastic neutron scattering. Physica B: Condensed Matter. 2018;**551**:291-296

[16] Vineyard GH. Scattering of slow neutrons by a liquid. Physics Review. 1958;**110**:999-1010

[17] de Gennes PG. Liquid dynamics and inelastic scattering of neutrons. Physica. 1959;**25**:825-839

[18] Wu B, Iwashita T, Egami T. Atomic dynamics in simple liquid: De Gennes

narrowing revisited. Physical Review Letters. 2018;**120**:135502

[19] Cargill GS III. Structure of metallic alloy glasses. Solid State Physics. 1975; **30**:227-320

[20] Sette F, Ruocco G, Krisch M, Masciovecchio C, Verbeni R, Bergmann U. Transition from normal to fast sound in liquid water. Physical Review Letters. 1996;**77**:83-86

[21] Scopigno T, Balucani U, Ruocco G, Sette F. Inelastic X-ray scattering and the high-frequency dynamics of disordered systems. Physica B. 2002; **318**:341-349

[22] Iwashita T, Nicholson DM, Egami T. Elementary excitations and crossover phenomenon in liquids. Physical Review Letters. 2013;**110**:205504

[23] Shinohara Y, Dmowski W, Iwashita T, Wu B, Ishikawa D, Baron AQR, et al. Viscosity and real space molecular motion of water: Observation with inelastic X-ray scattering. Physical Review E. 2018;**98**:022604

[24] Egami T, Shinohara Y. Dynamics of water in real space and time. Molecular Physics. 2019. DOI: 10.1080/00268976.2019.1649488. accepted for publication on May 8

[25] Shinohara Y, Dmowski W, Iwashita T, Ishikawa D, AQR B, Egami T. Local correlated motions in aqueous solution of sodium chloride. Physical Review Materials. 2019;**3**:065604

[26] Roseker W, Hruszkewycz SO, Lehmkühler F, Walther M, Shulte-Schrepping H, Lee S, et al. Towards ultrafast dynamics with split-pulse X-ray photon correlation spectroscopy at free electron laser sources. Nature Communications. 2018;**9**:1704

Chapter 4

Synchrotron Radiation-Based Quasi-Elastic Scattering Using Mössbauer Gamma Ray with neV-Energy Resolution

Makina Saito, Toshiji Kanaya and Ryo Mashita

Abstract

Gamma rays of energy 14.4 keV from excited ^{57}Fe nuclei show a very narrow energy width of 4.67 neV by the Mössbauer effect. Mössbauer gamma rays are utilised as probe beams in unique quasi-elastic scattering spectroscopy with neV-energy resolution. The technique enables measurements of atomic/molecular dynamics on timescales between nanoseconds and microseconds for various condensed matter systems, such as supercooled liquids, glasses and soft materials. The microscopic dynamics is measured in time domain or energy domain based on synchrotron radiation using a time-domain interferometer or a nuclear Bragg monochromator, respectively. We introduce state-of-the-art spectroscopic techniques, application results and future perspectives of quasi-elastic Mössbauer gamma ray scattering based on synchrotron radiation.

Keywords: Mössbauer gamma ray, synchrotron radiation, quasi-elastic scattering, glass transition, slow dynamics

1. Introduction

The recoilless nuclear excitation of a gamma ray and its reversal process of recoilless gamma ray emission were first reported by Mössbauer [1]. These phenomena occur in solids when the recoil momentum of gamma rays in absorption and emission processes is taken up by the whole crystal. Consistently, this physical phenomenon is referred to as the Mössbauer effect [2]. For ^{57}Fe nuclei, the excitation energy to the first excited state is 14.4 keV, whereas the uncertainty width of the excited state $\Gamma_0 \sim 4.67$ neV is relatively very narrow. Therefore, the gamma rays emitted from the excited ^{57}Fe nuclei by the Mössbauer effect show an energy $E_0 \sim 14.4$ keV and a natural energy width $\Gamma_0 \sim 4.67$ neV. The photon emitted by the nuclei is called the gamma ray because it originates at the nucleus. However, Mössbauer gamma rays have lower energy than gamma rays involved in astronomy physics and are, instead, closer to the energy range of hard X-rays. In this chapter, we refer to such gamma rays as Mössbauer gamma rays. In these cases, the ratio of the gamma rays' energy to the natural energy width reaches $\Gamma_0/E_0 \sim 10^{-13}$, indicating that the Mössbauer gamma rays exhibit very high monochromaticity. The surrounding electrons affect nuclear excitation energies through hyperfine interactions. Therefore, electronic states around the specific Mössbauer nuclei can be selectively studied from the measured nuclear excitation energies via the Mössbauer effect. This spectroscopic technique, known as Mössbauer spectroscopy, has been widely used for more than 40 elements and 70 nuclear species (referred to as the Mössbauer nuclear species) to resolve various challenges in the fields of chemistry, physics, geology and biology [2].

Microscopic dynamics in condensed matter, which do not contain Mössbauer nuclear species, have been studied since the 1960s with Mössbauer gamma rays [3]. In these experiments, the Mössbauer effect is utilised to generate the monochromatic gamma rays from a radioactive isotope (RI) source, and a quasi-elastic scattering experiment is performed for some samples [3]. In this chapter, we refer to the methods as quasi-elastic gamma ray scattering (QEGS) spectroscopy based on conventional nomenclature, such as inelastic/quasi-elastic neutron/X-ray scattering though this method has often been referred to as the Rayleigh-scattering of Mössbauer radiation method. The neV-energy resolution of the gamma rays from ^{57}Fe nuclei allows the dynamics to be measured on timescales of about 100 ns. However, the measurements require much longer times because gamma rays from RI sources do not have parallel beams with enough brilliance for the QEGS experiment.

Recently, synchrotron radiation (SR)-based QEGS spectroscopic techniques using a ^{57}Fe-nuclear Bragg monochromator (NBM) [4, 5] and a time-domain interferometer (TDI) of ^{57}Fe gamma rays [6] have been developed. These methods have enabled much faster measurements of the atomic/molecular dynamics than RI-based QEGS spectroscopy, owing to the high brilliance and directionality of the SR source. To date, alloys, supercooled molecular liquids, polymers, ionic liquid, liquid crystals and polymer nanocomposite systems have been studied by SR-based QEGS spectroscopy.

In this chapter, we consider Mössbauer gamma rays from ^{57}Fe nuclei because the gamma ray is most frequently used for QEGS spectroscopy. The length scales of the density correlation function currently observable by SR-based QEGS spectroscopy using TDI range from 0.1 to 6 nm, and the fluctuation timescales vary from few nanoseconds to sub-microseconds, as shown in **Figure 1**. The figure demonstrates how QEGS spectroscopy enables us to study density fluctuations, which are quite difficult to study by conventional spectroscopies in the microscopic range. Many unsolved issues are related to these time and length scales, including microscopic activation processes, which are related to the nature of the glass transition, start to occur in glass-forming materials in the time and length scales with cooling.

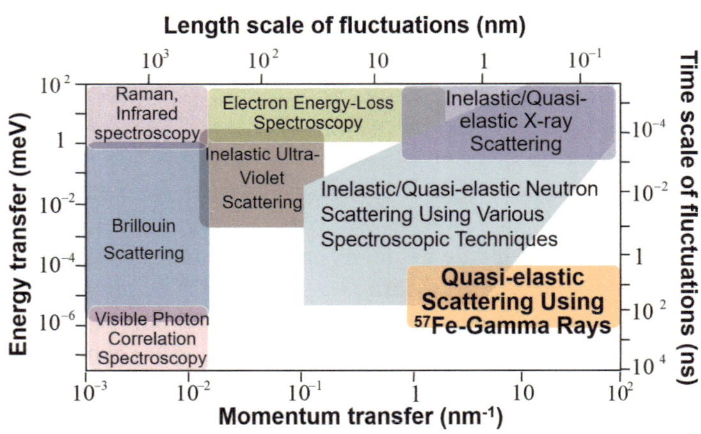

Figure 1.
Various experimental techniques and the covered time and length scales. Quasi-elastic scattering spectroscopy using gamma rays from ^{57}Fe covers a unique time- and length-scale region.

This chapter is organised as follows: In section 2, basic concepts of quasi-elastic scattering are introduced, and QEGS spectroscopic techniques are explained. In section 3, experimental results of application studies on several supercooled glass formers are described. In section 4, we conclude this chapter by describing future perspectives of QEGS.

2. Quasi-elastic scattering spectroscopy using Mössbauer gamma rays

In this section, we introduce the quasi-elastic scattering technique using Mössbauer gamma rays. In section 2.1, basic concepts of the quasi-elastic scattering technique are described. In section 2.2, we introduce energy-domain spectroscopic techniques of QEGS using Mössbauer gamma rays from conventional RI and SR sources. In section 2.3, time-domain measurement techniques of QEGS spectroscopy using single-line and multi-line TDI are described.

2.1 Introduction to quasi-elastic scattering

In this scattering process, gamma rays with wavevector $\boldsymbol{k}$ are emitted from the excited ^{57}Fe nuclei by the Mössbauer effect and Mössbauer gamma rays impinge on a sample. The geometry of the resulting Rayleigh-scattering process is shown in **Figure 2**, where $\boldsymbol{k}'$ is the wavevector of the scattered gamma rays and $\boldsymbol{q} = \boldsymbol{k}' - \boldsymbol{k}$ is the transferred momentum vector of the gamma rays to the sample [7]. The electron density field in the sample can be written as $\rho(\boldsymbol{r},t) = \sum_{i=1}^{N} \delta(\boldsymbol{r}-\boldsymbol{r}_i(t))$, where $\boldsymbol{r}$ and t are the space coordinate and the time, respectively, N is the molecular number in the sample and $\boldsymbol{r}_i$ is the centre position of atom i. In the momentum transfer (wavenumber) space, the density field $g(\boldsymbol{q},t)$ is written as $g(\boldsymbol{q},t) = \sum_{i=1}^{N} \exp\{i\boldsymbol{q} \cdot \boldsymbol{r}_i(t)\}$. Due to atomic/molecular motions in the sample, the gamma rays transfer energy to the sample and vice versa. In quasi-elastic scattering processes, a neV-energy broadening of the gamma rays energy is observed, as shown in **Figure 2**. This peak broadening is due to energy transfers that occur at neV-energies, which are thus much smaller than the incident gamma rays' energy, for which we can thus assume $|\boldsymbol{k}| \sim |\boldsymbol{k}'|$. Consequently, the amplitude of the transferred momentum is $q = 2\,|\boldsymbol{k}|\,\sin(\theta)$, where 2θ is the scattering angle. When the sample shows disordered structures, as in liquids and glasses, the relevant variable is the absolute value q rather than the vector $\boldsymbol{q}$.

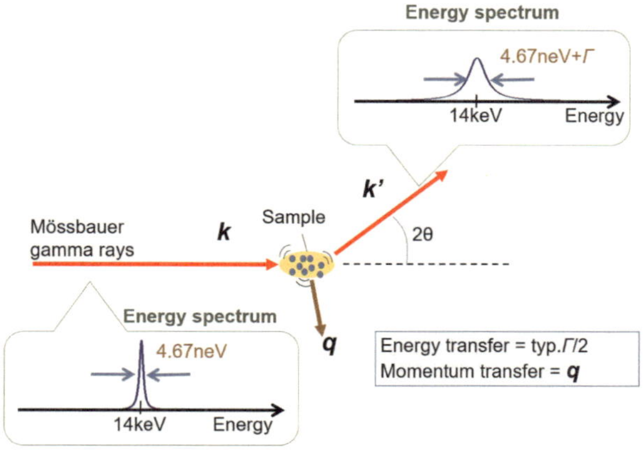

Figure 2.
Schematic picture of the quasi-elastic scattering process of Mössbauer gamma rays from a sample.

We introduce the spatial correlation function of the electron density $G(\boldsymbol{r})$ as $G(\boldsymbol{r}) = \langle \rho(\boldsymbol{r_0}+\boldsymbol{r}, t_0)\rho(\boldsymbol{r_0}, t_0)\rangle$, where $\langle \cdots \rangle$ denotes the equilibrium average over t_0 and position $\boldsymbol{r_0}$, and r is a distance. The static structure factor $S(\boldsymbol{q})$ is defined as its space Fourier transform $S(\boldsymbol{q}) = \int G(\boldsymbol{r}) \exp(\mathrm{i}\boldsymbol{q} \cdot \boldsymbol{r})d\boldsymbol{r}$. For simple monoatomic liquids, the scattering intensity $I(q)$ is related to $S(q)$ as $I(q) = N\,S(q)$. From this definition, it appears that the scattering at a given q is mainly caused by atomic pair correlations roughly occurring over distances $2\pi/q$, in a very simple picture. At atomic scales, $S(q)$ is obtained via X-ray and neutron diffraction experiments.

We introduce the time and space correlation function $G(\boldsymbol{r}, t) = \langle \rho(\boldsymbol{r_0}+\boldsymbol{r}, t_0 + t)\rho(\boldsymbol{r_0}, t_0)\rangle$ describing the microscopic structural dynamics. Its q-domain representation, often called the intermediate scattering function, is $S(\boldsymbol{q}, t) = \int G(\boldsymbol{r}, t) \exp(\mathrm{i}\boldsymbol{q} \cdot \boldsymbol{r})d\boldsymbol{r}$ and can be measured by neutron spin echo spectroscopy and photon correlation spectroscopy. The spectral intensity of the scattered gamma rays at a given q is $I(q, E) = NS(q, E)$, where $S(\boldsymbol{q}, E) = \int G(\boldsymbol{r}, t) \exp\left[\mathrm{i}(\boldsymbol{q} \cdot \boldsymbol{r} - tE/\hbar)\right]dtd\boldsymbol{r}$ is called the dynamics structure factor. Inelastic/quasi-elastic X-ray scattering using meV-high energy resolution monochromators and neutron scattering using triple-axis spectrometers measure $S(q, E)$. Both $S(q, E)$ and $S(q, t)$ show quantitatively equivalent information for $G(r, t)$.

2.2 Energy-domain spectroscopy of QEGS

In this section, we consider QEGS-based energy-domain spectroscopic techniques using Mössbauer gamma rays from conventional RI and SR sources. **Figure 3a** shows the common experimental design of the technique [8, 9]. In the setup, monochromatic Mössbauer gamma rays impinge on the sample. The quasi-elastic broadening of the scattered gamma ray's energy is analysed by the ^{57}Fe-Mössbauer absorber, as explained below. As **Figure 3b** shows, $S(q, E)$ is observed as a transmittance-type spectrum $I(q, E)$, which is conceptually written as $I(q, E) \propto 1 - \int dE' S(q, E')R(E - E')$, where $R(E)$ is the resolution function.

2.2.1 RI-based QEGS spectroscopy: Rayleigh-scattering Mössbauer radiation

Rayleigh-scattering Mössbauer radiation (RSMR) spectroscopy is a conventional QEGS spectroscopic technique that uses RI as the source of the gamma ray probe. RSMR spectroscopy has been used to study microscopic dynamics in glass formers, proteins and liquid crystals as summarised in a review by Champeney [8]. In this method, monochromatic Mössbauer gamma rays (e.g., from a radioactive ^{57}Co source with an energy E_0 of 14.4 keV and an energy width of 4.67 neV) are sent to the sample. A broadening of the energy width of the quasi-elastically scattered

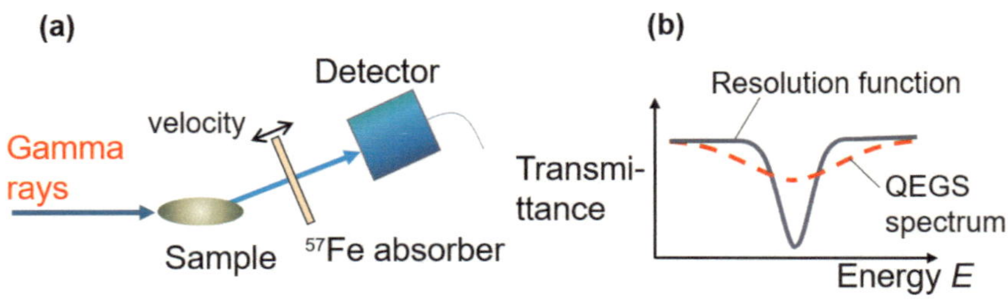

Figure 3.
Energy-domain QEGS experimental setup and typical spectrum. (a) Schematic figures of QEGS experimental setups for energy-domain measurement and (b) energy spectra of the resolution function (solid line) and QEGS energy spectra in the presence of measurable dynamics (dashed line).

gamma rays from a sample is detected by an absorption spectroscopy method commonly used in Mössbauer spectroscopy (**Figure 3a**). A transmittance-type energy spectrum is obtained by scanning the velocity v of a movable ^{57}Fe gamma ray absorber with a single-line excitation profile. The absorber acts as the energy analyser, since its velocity determines the relative energy shift $E = E_0 v/c$ via the Doppler effect, where c is the speed of light. RSMR measurements require ample measuring time (at least several weeks) to obtain a spectrum with enough statistics for analysis because the RI source emits gamma rays in all directions, and limited flux is introduced to the sample.

2.2.2 SR-based QEGS spectroscopy using ^{57}Fe-nuclear Bragg monochromator

The QEGS-based energy-domain spectroscopic technique using an SR source was developed with the ^{57}Fe-NBM [4, 5]. NBM is used for a specific condition, in which conventional X-ray diffraction by electrons is forbidden, while nuclear resonant diffraction with nuclear excitation and deexcitation processes is allowed. In such cases, we can detect almost pure Mössbauer gamma rays on a 10 neV-energy width scale due to the specific Bragg angle selectively from a very intense incident SR. Therefore, the SR-NBM system is often called as synchrotron Mössbauer source [10]. The SR-based QEGS experiment has higher efficiency than conventional RSMR using RI because the monochromatic gamma rays from the NBM exhibited high directivity [10]. Moreover, the energy width of the Mössbauer gamma ray probe could be controlled to be much larger than the natural-line width (i.e., up to μeV) [11]. This unique characteristic of SR-based QEGS spectroscopy using NBM allows us to measure microscopic dynamics up to sub-nanosecond timescales.

2.3 Time-domain measurement of QEGS

The time-domain spectroscopy of QEGS is achieved using TDI. In this section, we introduce time-domain spectroscopic techniques.

2.3.1 SR-based QEGS using single-line TDI

The measurement principles of QEGS using the simplest TDI (usually referred to as single-line TDI) are described here. We discuss TDI using Mössbauer gamma rays from ^{57}Fe because it exhibits the highest utility among nuclear species potentially available for TDI. **Figure 4a** shows the schematic experimental setup [6, 12, 13].

First, we consider the nuclear forward scattering (NFS) case, which often provides a calibration for the QEGS measurement because it is not affected by the dynamics of the sample. In the upper panel of **Figure 4a**, we show the experimental design for the NFS experiment using TDI. The incident SR crosses two identical materials with a single-line ^{57}Fe nuclear excitation profile corresponding to the nuclear time response function $G(t)$ ultimately detected by the detector. Most of the SR beam crosses the ^{57}Fe materials without any interaction. A small portion (typically $\sim 10^{-6}$) of SR excites the ^{57}Fe nuclei in the materials, causing the gamma rays to emit when the excited ^{57}Fe nuclei decay. The gamma rays travel undeflected towards the forward detector because of the high directivity inherited from the incident SR. The gamma rays can be distinguished from the much more intense SR because they are delayed from the SR pulse by a typical delay time coincident with the lifetime of the excited ^{57}Fe nuclei ($\sim$100 ns). The upstream material is moved with a constant velocity to change the relative nuclear excitation energy ΔE through the Doppler effect and consequently the energy spectrum of the gamma rays at the detector position shows two peaks due to the difference in the gamma ray energy

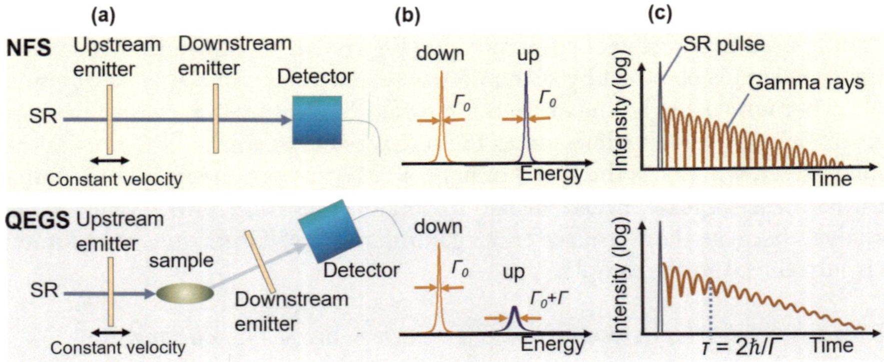

Figure 4.
NFS and QEGS using single-line gamma rays TDI. (a) Experimental setups and examples of (b) energy spectra and (c) time spectra for NFS and QEGS using single-line TDI.

between the two materials (see the upper panel of **Figure 4b**). The time resolution of the detector is typically 1 ns, which is much shorter than the lifetime of excited ^{57}Fe, which enables to measure the time spectrum of the delayed gamma rays with high precision. The upper panel of **Figure 4c** shows the corresponding time spectrum. We can see the decay of the gamma rays' intensity on the timescale of excited ^{57}Fe. On the time spectrum, there is a beating pattern caused by the interference of the gamma rays with two peaks in the energy spectrum.

Next, we consider the QEGS case, corresponding to the scattering of the sample at a finite angle. In the lower panel of **Figure 4a**, we show the QEGS experimental design. The incident SR is scattered by a sample and detected by the detector. Two identical materials with a single-line ^{57}Fe nuclear excitation profile are placed on the beam path in front of and behind the sample. This system is called the single-line TDI because each material that emits gamma rays (here, referred to as single-line emitter) shows a single-line nuclear excitation profile. A typical energy spectrum of gamma rays at the detector position is shown in the lower panel of **Figure 4b**. The gamma rays from the upstream emitter (denoted as 'up' in **Figure 4b**) are quasi-elastically scattered by the sample and the energy width is broadened as Γ. However, the energy width of the gamma rays from the downstream emitter (denoted as 'down' in **Figure 4b**) is not broadened because it is emitted by the sample after the scattering process.

Next, we considered the time spectrum of the gamma rays obtained by the detector for the QEGS case. When the energy shift is sufficiently large ($\Delta E \gg \Gamma_0$and $\Delta E \gg \Gamma$), the radiative coupling effect can be neglected [6, 12, 13]. Additionally, we can assume that the incident SR showed a temporal pulse structure with negligible width. In such cases, the electric field $E(q,t)$ at detector position at an angle corresponding to q can be written as

$$E(q,t) \propto \delta(t) + g(q,t)G(t)\mathrm{e}^{\mathrm{i}\omega_{\mathrm{Q}}t} + g(q,0)G(t) \tag{1}$$

where $\omega_{\mathrm{Q}} = \Delta E/\hbar$ is the angular frequency of the beating pattern. We ignored the coefficient of the transmittance because it does not affect the final spectrum shape. The first, second and third terms of Eq. (1) represent the electric field amplitudes of the prompt SR, gamma rays emitted from the upstream and downstream emitters, respectively. The delayed gamma rays' measurement for part of the obtained time spectrum $I(q,t)$ is written as

$$I(q,t) \propto |G(t)|^2 S(q)\left[1 + S'(q,\mathrm{t})\cos\left(\omega_{\mathrm{Q}}t\right)\right]. \tag{2}$$

In Eq. (2), $S'(q,t)$ is the intermediate scattering function normalised by $S(q)$, that is, $S'(q,t) \equiv S(q,t)/S(q)$ and the static structure factor $S(q)$ is $S(q) = \left\langle |g(q,t)|^2 \right\rangle = \left\langle |g(q,0)|^2 \right\rangle$ [6, 12, 13]. We assume that the scattering from a sample with a macroscopic number of atoms was measured with an acquisition time long enough to provide a reliable determination of the relevant average ensembles. For an NFS experiment under the same emitter conditions, the NFS time spectrum is expressed by Eq. (2) with $S(q) = 1$ and $S'(q,t) = 1$. Examples of time spectra for NFS and QEGS cases with a relaxation time τ of 100 ns are shown in the upper and lower panels of **Figure 4c**, respectively. For the actual fitting of the spectra, the time resolution of the detector and constant background noise would need to be considered.

Next, we considered the meaning of the time spectrum. The broadening of the gamma rays by an energy width Γ reflects the dynamics in a sample. The broadening induces the distribution of the beat frequency in a time domain and this effect is seen as the relaxation of the beating pattern with the relaxation time $\tau = 2\hbar/\Gamma$ in the simplest case. Further consideration revealed that the relaxation time of the beating pattern coincides with the relaxation time of the density correlation in the sample (namely, the intermediate scattering function) [6, 12, 13]. This analysis is a basic interpretation of how the time spectrum reflects the dynamics in a sample. We note that an intrinsic relaxation of $S'(q,t)$ caused by an external vibration, for example, should also be considered for the actual dynamics study.

2.3.2 SR-based QEGS spectroscopy using multi-line TDI

Here, we consider QEGS spectroscopy using multi-line TDI [14]. In this case, emitters with several nuclear excitation energies are used for TDI. We assume again that the two emitters show different excitation energies from each other. Generally, the nuclear time response functions in emitters are different from each other in multi-line cases. Therefore, we introduce the time response functions for the upstream and downstream emitters as $G_1(t)$ and $G_2(t)$, respectively. In such cases, we obtain the expression $E(q,t)$ as $E(q,t) \propto \delta(t) + g(q,t)G_1(t) + g(q,0)G_2(t)$ from Eq. (1). The intensity of the delayed gamma rays can be written as

$$I(q,t) \propto S(q)\left\{|G_1(t)|^2 + |G_2(t)|^2 + S'(q,t)\left[G_1^*(t)G_2(t) + G_1(t)G_2^*(t)\right]\right\}. \quad (3)$$

As an example of multi-line TDI, we considered α-iron foils as emitters, where the nuclear excitations are allowed for six different energies without an external magnetic field. **Figure 5a** shows an experimental setup using α-iron emitters. When the magnetic field is applied to the α-iron foils, as shown in **Figure 5a**, the transitions allowed in the two emitters are selected to be different from each other. Consequently, the gamma rays' energy emitted from these two emitters is different, as shown in **Figure 5b**, where the gamma rays from the upstream and downstream emitters are denoted as 'up' and 'down', respectively. Examples of the energy spectra of gamma rays for cases without atomic motion and motion with a relaxation time of 100 ns are shown. **Figure 5c** depicts the corresponding time spectra. The beating pattern changes following the decay of $S'(q,t)$. By introducing the multi-line condition, the interference beating pattern of the gamma rays on the time spectrum becomes more complex than the single-line case. However, the incident SR can be more effectively utilised for experiments and the gamma rays' count rate increases. Additionally, it can be shown that the time spectrum changes more drastically, reflecting the dynamics [14]. These properties of the multi-line TDI greatly improve the measurement efficiency in comparison to the single-line method.

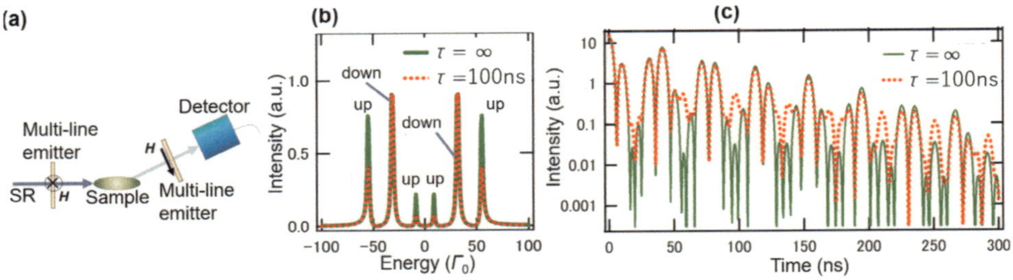

Figure 5.
Examples of quasi-elastic scattering using multi-line gamma rays TDI. (a) Experimental setup and examples of (b) energy spectra and (c) time spectra in the cases of no relaxation and $\tau = 100$ *ns.*

2.3.3 SR-based QEGS using TDI considering energy resolution of incident SR

Here, we consider the effect of the energy width of the incident SR on the gamma rays' time spectrum obtained by the QEGS experiment. After the first induced heat load from the Si(111) monochromator, the SR showed a relatively broad energy profile; an energy width of the eV order could be considered white for the QEGS system. However, the incident SR is usually further monochromatised by using a high-resolution monochromator (HRM). This device generates typical energy widths in the meV range to suppress radiation damage to the system [6, 12–14]. The meV-energy interval is equivalent to or smaller than the energy scale of phonons in samples. Therefore, a portion of the incident SR transfers a larger amount of energy to the sample by interacting with the phonons. We found that the inelastic scattering process affects the intensity ratio of the gamma rays from the upstream and downstream emitters. Considering this effect, we modify Eq. (3) as

$$I(q,t) \propto S(q)\left\{|G_1(t)|^2 + |G_2(t)|^2 + S'(q,t)[G_1^*(t)G_2(t) + G_1(t)G_2^*(t)] - (1 - f_{\Delta E})|G_2(t)|^2\right\}, \tag{4}$$

where $f_{\Delta E}$ is the factor reflecting the sample dynamics on a meV-energy scale [14]. It was confirmed that the QEGS time spectrum obtained using TDI with multi-line gamma rays could be nicely analysed using Eq. (4) [14]. Additionally, we showed that QEGS spectroscopy using HRM originally has two resolution functions on neV- and meV-energy scales. By using multi-line TDI in the condition $|G_1(t)|^2 \neq |G_2(t)|^2$, dynamical information, such as the elastic scattering intensity, can be obtained simultaneously on nanosecond and sub picosecond timescales [14].

3. Application results of SR-based QEGS using TDI

To date, SR-based QEGS spectroscopy has been used to study glass-forming molecular liquids [15–19], polymers [20], polymer nanocomposites [21], ionic liquids [22], alloys [23] and liquid crystals [24].

3.1 Microscopic dynamics in glass formers

The general mechanism of the liquid-glass transition phenomenon, which has not been revealed, has attracted much interest. It is widely accepted that a relaxation process, known as the α process, is closely related to glass transitions [25–27].

Therefore, atomic and molecular dynamics of supercooled glass formers have been energetically investigated to understand glass transitions. The temperature (T) dependencies of the α-relaxation time and viscosity of some glass formers show super-Arrhenius behaviour when cooled towards the glass transition temperature [28]. These behaviours are often fitted by the Vogel-Fulcher-Tammann (VFT) law: $\eta(T), \tau(T) \propto \exp\left[DT_0/(T - T_0)\right]$, where D is the fragility index and T_0 is the Vogel-Fulcher temperature [25–27]. The slope of this super-Arrhenius behaviour is determined by the fragility index and its physical origin still has not been fully elucidated; this limitation remains one of the central challenges in studying glass transitions.

The other challenging task in these systems is understanding the origin of the dynamical change of the α process, which starts to occur at a temperature of $\sim$1.2 T_g upon cooling, where T_g is the glass transition temperature. The changing temperature is recognised as the dynamical crossover temperature T_c [29]. In addition to the α process, various processes have been observed in relaxation maps, which summarise the temperature dependence of processes in glass formers. Among the various relaxation processes subjected to a thorough scrutiny, it is worth mentioning the Johari-Goldstein (JG)-β process, which emanates from the α process in relaxation maps and, instead, follows Arrhenius behaviour $\tau(T) \propto \exp\left[E_A/(RT)\right]$ even below the glass transition temperature, where E_A is the activation energy and R is the gas constant [30]. Recently, the JG-β process was believed to commonly exist in supercooled glass formers and relate to the nature of the glass transition mechanism [27]. The branching temperature of the JG-β process from the α process $T_{\alpha\beta}$ is frequently seen near the dynamical crossover temperature T_c. This synchronism is believed to be an intrinsic feature of supercooled glass formers. However, the dynamical crossover and branching phenomena are far from being understood fully. Conventional methods, such as dielectric relaxation spectroscopy, do not provide spatial-scale information on the dynamics, and the α and JG-β processes are not clearly discerned around T_c and $T_{\alpha\beta}$. Therefore, $T_{\alpha\beta}$ has been estimated as a crossing point of the α-relaxation time and an extension of the JG-β relaxation time by assuming the Arrhenius law [27].

Understanding the microscopic dynamics around T_c and $T_{\alpha\beta}$ is indispensable to elucidating the glass transition mechanism. SR-based QEGS spectroscopy is a method ideally suited to understand the microscopic dynamics in deeply supercooled glass formers around T_c and $T_{\alpha\beta}$ and its evolution towards the glass transition. This technique enables to measure the atomic/molecular dynamics with specification of its spatial scale on a nanosecond/microsecond timescale, where the JG-β process commonly occurs [27]. We performed SR-based QEGS experiments using single-line and multi-line TDI on various glass formers. We introduce the results on *o*-terphenyl in section 3.2 and polybutadiene in section 3.3. Additionally, the application results of a polymer nanocomposite system are discussed in section 3.4.

3.2 Results on *o*-terphenyl

o-terphenyl (OTP) is widely studied as a model system of glass formers. For deeply supercooled OTP, it has been reported that a change in the VFT parameters of the α-relaxation timescale occurs at around 290 K [31]. Additionally, decoupling of the rotational and translational diffusion coefficients of the tracer molecules [32], an abrupt decrease of the stretching parameter of the α-relaxation form [33] and formation of a cusp in the temperature dependence of f_{DW} [34] have been reported at 290 K upon cooling. From these observations, 290 K can be recognised as the dynamical crossover temperature T_c. Additionally, dielectric relaxation spectroscopy studies have revealed that the branching temperature $T_{\alpha\beta}$ ~290 K is almost the same as T_c for OTP [35]. Again, $T_{\alpha\beta}$ should be carefully interpreted because it was estimated by the extrapolation assumption whose validity has not been confirmed. Moreover, the details of the branching phenomenon of the α and JG-β processes and the relationship between $T_{\alpha\beta}$ and T_c are still unclear.

We studied OTP using single-line ^{57}Fe gamma rays TDI for the QEGS measurements [16, 20]. Detectors were placed at angles corresponding to q values shown in the inset of **Figure 6**. For analysing the time spectra, we applied Eq. (2) assuming the Kohlrausch-Williams-Watts (KWW) function $f \exp\left[-(t/\tau)^{\beta_{\mathrm{KWW}}}\right]$ for the relaxation of $S'(q,t)$, where f is the amplitude of the relaxation, τ is the relaxation time, and β_{KWW} is the stretching parameter. In the obtained QEGS time spectra, the values of β_{KWW} were determined to be 0.56 (±0.31) for q values ranging from 27 to 50 nm^{-1} (265 K), 0.64 (±0.15) for 14 nm^{-1} (285 K) and 0.66 (±0.25) for 16–31 nm^{-1} (285 K) [16]. Therefore, we analysed all QEGS time spectra by using a fixed β_{KWW} of 0.6 [34]. The τ obtained by fitting was transformed to a mean relaxation time $\langle\tau\rangle$ using the equation $\langle\tau\rangle = \tau\,\Gamma(1/\beta_{\mathrm{KWW}})/\beta_{\mathrm{KWW}}$, where Γ is the gamma function. In **Figure 6**, we show the obtained relaxation map with spatial-scale information of the dynamics as a q dependence.

Figure 6 depicts the temperature dependence of $\langle\tau\rangle$. At q = 14 nm^{-1}, the temperature dependence obeys the VFT law, as suggested by the comparison with best-fitting curve obtained by the least-squares method. The best-fitting VFT parameters were determined to be D = 3.1(±0.3) and T_0 = 235(±26) K. We confirmed that the obtained $\langle\tau\rangle$ values are consistent with reported α-relaxation times [35]. This agreement suggests that the relaxation observed at q = 14 nm^{-1} is caused by the α process.

In the larger q region (18 < q < 42 nm^{-1}), the temperature dependencies of $\langle\tau\rangle$ were found to obey the VFT law above 278 K, as shown in **Figure 6**; however, these data follow the Arrhenius law below 278 K (1000/T ~ 3.6 K^{-1}). The turning temperature of 278 K appears to be independent of q within experimental error. The obtained $\langle\tau\rangle$ values at q = 18 nm^{-1} are similar to the JG-β relaxation times obtained by dielectric relaxation spectroscopy (shown as a dashed line in **Figure 6**) [31]. Further consideration revealed that the relaxations observed at the large q region

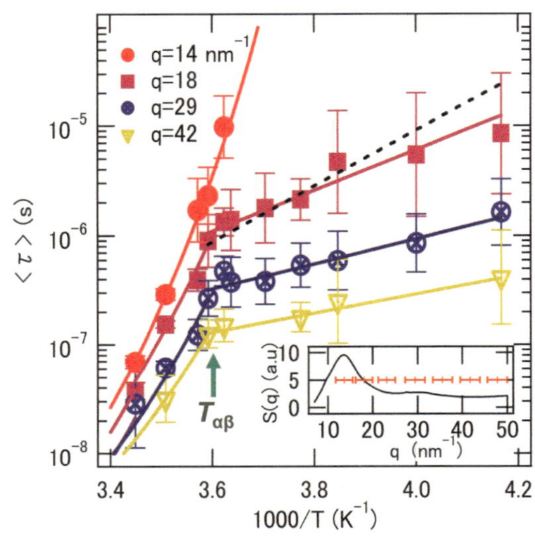

Figure 6.
Temperature dependence of $\langle\tau\rangle$ at typical q *values. The solid lines are curves fitted by the Arrhenius and VFT laws. The inset shows S(q) and corresponding q values of the measurement. The dashed line shows the JG-β relaxation time obtained by dielectric relaxation spectroscopy.*

below 278 K are mainly caused by the JG-β process [16]. Since the T dependence of $\langle\tau\rangle$ above 278 K at q = 14 nm^{-1} is similar to the one observed at larger q values, we deduce that above 278 K, α-relaxations dominate. Therefore, the temperature of 278 K can be identified as the branching temperature $T_{\alpha\beta}$, at which, at sufficiently large q's, the main relaxation changes from the α to JG-β type upon cooling. Furthermore, since JG-β emerges only at large q values ($18 < q < 42$ nm^{-1}), these have a mainly local character [36–38]. The obtained branching temperate $T_{\alpha\beta}$~278 K is clearly different from the one, $T_{\alpha\beta}$~290 K, previously obtained from the extrapolation of the JG-β relaxation time. This discrepancy suggests that this extrapolation method is not appropriate as also shown in Ref. [16]. Additionally, we found a clear correlation in which: T_c~ 290 K> $T_{\alpha\beta}$ ~278 K. This relation suggests that the branching of the JG-β process requires further cooling from the onset temperature of the α process dynamical change, which is in contrast to previous reports.

3.3 Results on polybutadiene

As mentioned in section 3.2, the nature of glass transition is still not fully understood despite thorough investigative efforts [39]. In the last three decades, extensive studies on glass transitions have been performed theoretically, experimentally and by computer simulations. One of the most important experimental results constructed relaxation time maps of several glass-forming materials [35] by predicting the decoupling of the JG-β process from the α process. Extensive experimental studies have been performed to reveal the decoupling mechanism using various techniques such as NMR [40], dielectric relaxation (DR) [41] and neutron spin echo (NSE) [42–44]. We performed QEGS measurements using single-line TDI on polybutadiene (PB), which is a typical glass-forming polymer, to decouple the JG-β process from the α process [20].

The sample used in this experiment was 1,4-cis-trans-polybutadiene (PB), which is never crystallised because of the microstructure of cis:trans:vinyl = 47:46:7. The T_g determined by differential scanning calorimetry (DSC) was 170 K. The same QEGS measurements were performed on PB as OTP, and the observed time spectra were analysed in the same manner as OTP using Eq. (2). In the analyses, we employed a stretched exponential function with an exponent β_{KWW} of 0.45 as an intermediate scattering function [42–44].

The average relaxation time $\langle\tau_{KWW}\rangle$ obtained from the fitting curve is shown in **Figure 7** as a function of the inverse of absolute temperature 1/T at q = 9.6, 15, 21, 27, 32 and 39 nm^{-1}. The average relaxation times $\langle\tau_{KWW}\rangle$ evaluated from the fitting curve are consistent with those reported in the literature [20, 40–44]. The viscosity timescale $\tau_\eta(T)$ of PB is shown as a thick dashed line in **Figure 7**. This timescale was derived from the viscosity of the polymer melts via the Rouse model, which connects the viscosity η with the monomeric friction coefficient ζ through $\tau_\eta(T) \sim \zeta(T)/T$ [43]. The temperature dependence here is described by the VFT law. The temperature of $\langle\tau_{KWW}\rangle$ at q = 9.6 and 15 nm^{-1}, that is, slightly below and at the first peak in $S(q)$, shows very similar temperature dependence (the VFT law) to the viscosity timescale $\tau_\eta(T)$. Therefore, the VFT law was fitted to the observed $\langle\tau_{KWW}\rangle$ at q = 9.6 and 15 nm^{-1}. The lines in the figure are the results of the fits at various q values. The fact that the relaxation times observed at q = 9.6 and 15 nm^{-1} follow the same temperature dependencies as the viscosity timescale suggests that $\langle\tau_{KWW}\rangle$ at 9.6 and 15 nm^{-1} is dominated by the α process. On the other hand, the temperature dependencies of $\langle\tau_{KWW}\rangle$ at q = 21, 27, 32 and 39 nm^{-1} above the valley in $S(q)$ are very different from those at 9.6 and 15 nm^{-1}. At temperatures above ~210 K ($T_{\alpha\beta}$), it follows the VFT law, whereas it changes to Arrhenius behaviour

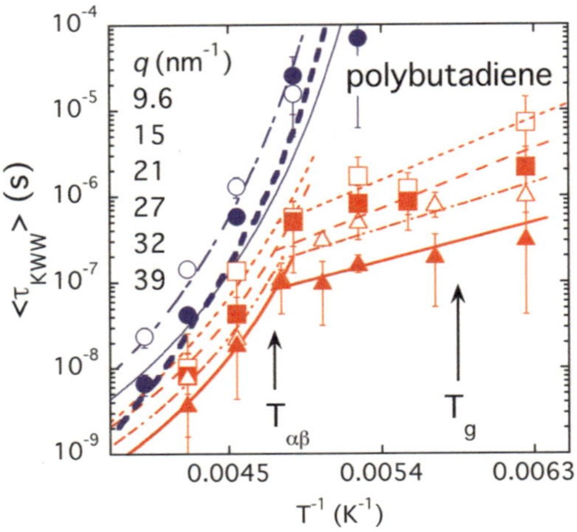

Figure 7.
Temperature dependence of the mean relaxation time $\langle \tau_{KWW} \rangle$ *obtained for PB at q = 9.6 (○, chain line), 15 (●, thin solid line), 21 (□, dotted line), 27 (■, dashed line), 32 (Δ, two-dot chain line) and 39 (▲, thick solid line) nm and fitting curves given by symbols and lines, respectively. Thick dashed line represents temperature dependence of viscosity timescale* $\tau_\eta(T)$.

below $T_{\alpha\beta}$. Here $T_{\alpha\beta}$ denotes the transition temperature. Therefore, the VFT law was used to fit data above $T_{\alpha\beta}$ at each q, whereas the Arrhenius equation was used for data below $T_{\alpha\beta}$. We thus conclude that the α process occurs above $T_{\alpha\beta}$, as suggested by the corresponding VFT behaviour, and it changes to the JG-β process below $T_{\alpha\beta}$, similar to what was observed in OTP. The observation here that the α process changes to the JG-β process at $T_{\alpha\beta}$ above the first peak in $S(q)$ contradicts the NSE results on PB [44] where no transition was observed in the high q range. It should be emphasised that this new transition finding at the higher q range can be attributed to the appropriate time and spatial resolutions of the SR-based QEGS technique for observing the branching phenomenon.

An extended mode coupling theory (eMCT) has been proposed to account for hopping processes [45]. This theory predicts a dynamical transition from the α process to a local, hopping-dominated, relaxation process at T_c. In other words, this transition corresponds to the switch of the temperature dependence from the VFT law to the Arrhenius law. In the eMCT framework, the transition from the α process to the JG-β process corresponds to the transition from the hydrodynamic continuous motion to the hopping motion. The fact that the transition above the first peak occurs near T_c supports this interpretation. In the present experiment, however, we observed that the α process persisted even below $T_{\alpha\beta}$ ($\sim T_c$) near the first S(q) peak. In other words, no transition occurred near the first peak. In this sense, the eMCT cannot be directly applied to our results.

The question still remains as to why the α process lasts even below $T_{\alpha\beta}$ at the first peak. Richter et al. have intensively studied relaxation processes in PB using NSE [42, 43] and found that the α process was observed at the first peak in $S(q)$, whereas the JG-β process was observed at the valley in $S(q)$ as mentioned above. These results agree with our observations below $T_{\alpha\beta}$. The key point of their results is that the intermolecular interaction is very important for understanding the transition. It has been demonstrated that the first peak in $S(q)$, the intermolecular correlation, is the strongest, leading to cooperative motion. However, at the valley, the intermolecular correlations are weaker than the first peak, and molecules move less cooperatively or freely. Hence, the cooperative α process is dominant at the first peak, and the isolated motion or the slow JG-β process is dominant at the valley. According to the eMCT, the α-relaxation changes to the JG-β one (hopping process) at around T_c even at the first peak of $S(q)$; however, in our experiment, the α process was still observed below T_c ($\sim T_{\alpha\beta}$) at the first peak. This discrepancy must be due to the strong cooperativity of the relaxation over distances roughly corresponding to the first $S(q)$ peak, which was not included in the eMCT.

3.4 Results on polybutadiene with nano-silica

Tyre rubber has been continuously developed to improve various aspects of its performance, such as its grip, fuel consumption and wear resistance, by adding fillers such as silica nanoparticles and cross-linking agents [46, 47]. However, the microscopic mechanisms behind these improvements are still not fully elucidated and a better understanding is needed to further improve tyre products. Many studies have shown that confined polymer layers around nanoparticles affect the rubber's macroscopic properties [48–57]. Molecular-scale dynamics studies have also revealed that the presence of nanoparticles slows down the microscopic segmental α-relaxation motion and increases its heterogeneity [52, 53]. However, we still do not have a complete picture of the microscopic dynamics for these systems. Additionally, the effect of the particle size on the microscopic dynamics has not been elucidated.

To elucidate the effect of nanoparticles on the microscopic α-relaxation dynamics of polymers, we studied the microscopic dynamics of a polybutadiene (PB) and silica nanoparticle mixture by SR-based QEGS using multi-line TDI. Two types of samples were used for this experiment: pure 1,4-PB and 1,4-PB nanocomposites with silica nanoparticles. Two PB nanocomposites, PB-silica20 and PB-silica100, were prepared with 20 vol% of silica nanoparticles with average diameters of 20 and 100 nm, respectively. The glass transition temperature T_g of pure PB was determined to be ~180 K and no T_g difference could be detected among the three samples.

Figure 8 shows the obtained wide-angle X-ray scattering (WAXS) profile of the two nanoparticle samples. From these WAXS results, we confirmed that the position of the main peak, mainly reflecting the intermolecular correlation of the PB, had changed very little and was covered by the q region in the quasi-elastic scattering measurements (see the bar in the figure). Least-squares fits were performed for the obtained PB time spectrum using Eq. (4) modelling the normalised intermediate scattering function with a KWW profile. The value of β_{KWW} for pure PB was determined to be 0.48 ± 0.10 at $q \sim 14\ \mathrm{nm}^{-1}$, which is consistent with the previously reported β_{KWW} value of 0.45 [42]. We obtained τ by setting β_{KWW} to be 0.45 for the pure PB spectra and then calculated the mean relaxation time $\langle\tau\rangle$ from $\langle\tau\rangle = \tau\,\Gamma(1/\beta_{\mathrm{KWW}})/\beta_{\mathrm{KWW}}$, where Γ is the gamma function [42].

Next, for the PB nanocomposites with silica nanoparticles, the polymer dynamics was studied through the analysis of the relaxation time extracted from the intermediate scattering function, while also considering its non-relaxing component originating from the stable nanoparticles. For the polymer nanocomposite systems, it is known that the contribution of the α-relaxation of polymers to the intermediate scattering function can be treated as a KWW function [48, 49]. Therefore, we used the function $F(q,t) = f(q)\exp\left\{-[t/\tau(q)]^{\beta_{\mathrm{KWW}}(q)}\right\} + c(q)$ to fit the normalised intermediate scattering function for the time spectra of PB-silica100 and PB-silica20, where $c(q)$ is the contribution of the non-relaxing component. By fitting the time spectra obtained for PB-silica100 at 250 K, we determined that the contribution of the non-relaxing component was $c = 0.22 \pm 0.07$ at $q \sim 14\ \mathrm{nm}^{-1}$, assuming $\beta_{\mathrm{KWW}} = 0.45$. We used these values to analyse the time spectra of both PB-silica20 and PB-silica100 because (i) the volume fraction of silica nanoparticles was the

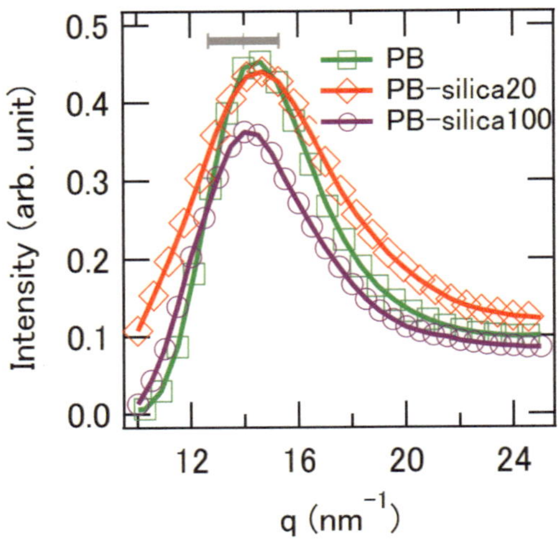

Figure 8.
WAXS profile obtained for pure PB and PB-Si100 at room temperature, and for PB-silica20 at 270 K. The bar represents the q region of the quasi-elastic scattering measurements.

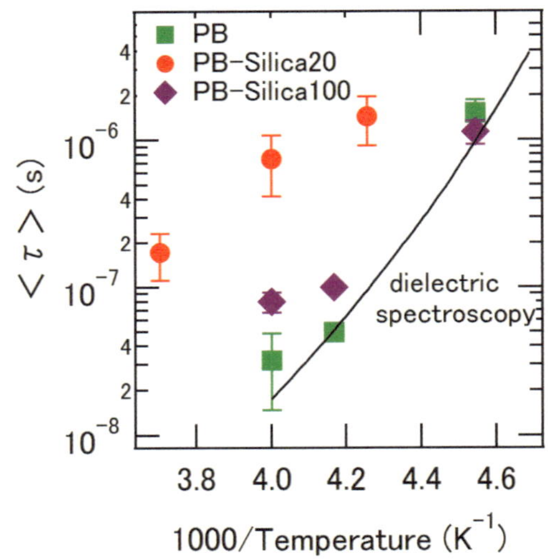

Figure 9.
Temperature dependence of the averaged relaxation times obtained for pure PB, PB-silica20 and PB-silica100 at q = 14 nm^{-1}. The error bars indicate the standard deviations, and the line indicates the α-relaxation times obtained by dielectric relaxation spectroscopy [58].

same for both samples and the c value could also be assumed to be the same and (ii) the non-relaxing component of the polymer was found to be negligible in a mixture of PB and carbon black nanoparticles in the q range of the first peak [48, 49].

Figure 9 shows the temperature dependence obtained for $\langle \tau \rangle$. The α-relaxation times of pure PB obtained by dielectric relaxation spectroscopy (depicted as a line in **Figure 9**) demonstrate that our results are consistent with the dielectric relaxation spectroscopy results [58]. The temperature dependencies of $\langle \tau \rangle$ obtained for PB-silica20 and PB-silica100 also show divergent behaviour, although the VFT parameters appear to be different compared to pure PB. At 250 K, the α-relaxation times obtained at q = 14 nm^{-1} for PB-silica20 and PB-silica100 were longer than those for pure PB, and this relation holds true throughout the studied temperature region. These data suggest that the nanoparticles cause the polymer α-relaxation motion to slow down. Moreover, the dynamics of PB-silica20 were much slower than

PB-silica100. Here, the volume fractions of silica nanoparticles in the PB-silica20 and PB-silica100 nanocomposites were the same, but the PB-silica20 surface area was on average 25 times larger than the PB-silica100 surface area. Therefore, the obtained results suggest that the polymer α-relaxation dynamics was restricted by contact with the surfaces of the nanoparticles and became even more restricted as the surface area increased. This result is consistent with the conventional idea that the α-relaxation times of polymers slow down due to interactions (chemical attachment and physical absorption) between the polymer and the silica nanoparticles on the surface [46, 47]. Additionally, these results demonstrate that QEGS can be used to reveal the polymer dynamics in nanocomposites and for characterising their microscopic dynamics; these insights will be important for advancing industrial materials such as tyre rubber. In the future, investigating the confinement effects of surface polymers/silica nanoparticles that are more similar to industrial tyre rubber will yield more specific information about improving tyre performance. The details of this work can be found in Ref. [21].

4. Conclusions and perspectives

Quasi-elastic scattering techniques using Mössbauer gamma rays are promising approaches for revealing nanosecond and microsecond dynamics directly from the microscopic viewpoint. Currently, quasi-elastic scattering systems using the gamma rays TDI have been developed and utilised for application studies. Additionally, by using a band-width variable ^{57}Fe-NBMs, we expect that the timescale of measurable dynamics will be expanded (e.g., up to sub 100 pico-second). Developing techniques that expand the timescales of measurements (i.e., between sub 100 pico-seconds and sub-microseconds), such as energy-domain quasi-elastic scattering systems combined with time-domain quasi-elastic scattering systems, is highly desirable.

Moreover, various new X-ray-based techniques are proposed for studying microscopic dynamics, based on focusing monochromators [59], or X-ray echo spectroscopy [60] or free electron lasers (e.g., four-wave mixing experiments) [61]. The combination of these new X-rays (and gamma rays)-based techniques expands the timescales of the measurements significantly (e.g., from femtoseconds to microseconds). Future studies will open new methodologies for depicting the microscopic structural dynamics of condensed matter by X-rays.

Author details

Makina Saito[1*], Toshiji Kanaya[2] and Ryo Mashita[3]

1 Institute for Integrated Radiation and Nuclear Science, Kyoto University, Kumatori, Japan

2 J-PARC, Material and Life Science Division, Institute of Material Structure Science, High Energy Accelerator Research Organization (KEK), Tokai, Japan

3 Sumitomo Rubber Industries, Ltd., Kobe, Japan

*Address all correspondence to: msaito@rri.kyoto-u.ac.jp

References

[1] Mössbauer RL. Kernresonanzfluoreszenz von gammastrahlung in Ir^{191}. Zeitschrift für Physik. 1958;**151**:124-143. DOI: 10.1007/BF01344210

[2] Greenwood NN, Gibb TC. Mössbauer Spectroscopy. London: Chapman and Hall Ltd.; 1971. 659 p. DOI: 10.1007/978-94-009-5697-1

[3] Elliott JA, Hall HE, Bunbury DSP. Study of liquid diffusion by Mössbauer absorption and Rayleigh scattering. Proceedings of the Physical Society. 1966;**89**:595-612. DOI: 10.1088/0370-1328/89/3/315

[4] Tischler JZ, Larson BC, Boatner LA, Alp EE, Mooney T. Time-sliced Mössbauer absorption spectroscopy using synchrotron radiation and a resonant Bragg monochromator. Journal of Applied Physics. 1996;**79**:3686-3690. DOI: 10.1063/1.361199

[5] Masuda R, Mitsui T, Kobayashi Y, Higashitaniguchi S, Seto M. A spectrometer for Rayleigh scattering of Mössbauer radiation using synchrotron radiation. Japanese Journal of Applied Physics. 2009;**48**:120221. DOI: 10.1143/JJAP.48.120221

[6] Baron AQR, Franz RH, Meyer A, Rüffer R, Chumakov AI, Burkel E, et al. Quasielastic scattering of synchrotron radiation by time domain interferometry. Physical Review Letters. 1997;**79**:2823-2826. DOI: 10.1103/PhysRevLett.79.2823

[7] Balucani U, Zoppi M. Dynamics of the Liquid State. Oxford: Oxford University Press; 1994. 336 p

[8] Champeney DC, Woodhams FWD. Investigation of molecular motions in supercooled liquids by Mössbauer scattering. The Journal of Physics B. 1968;**1**:620-631. DOI: 10.1088/0022-3700/1/4/313

[9] Mössbauer RL. Gamma-resonance and X-ray investigations of slow motions in macromolecular systems. Hyperfine Interactions. 1987;**33**:199-222. DOI: 10.1007/BF02394109

[10] Smirnov GV, van Bürck U, Chumakov AI, Baron AQR, Rüffer R. Synchrotron Mössbauer source. Physical Review B. 1997;**55**:5811-5815. DOI: 10.1103/PhysRevB.55.5811

[11] Mitsui T, Masuda R, Seto M, Hirao N. Variable-bandwidth ^{57}Fe synchrotron Mössbauer source. Journal of the Physical Society of Japan. 2018;**87**: 093001. DOI: 10.7566/JPSJ.87.093001

[12] Smirnov GV, Kohn VG, Petry W. Dynamics of electron density in a medium revealed by Mössbauer time-domain interferometry. Physical Review B. 2001;**63**:144303. DOI: 10.1103/PhysRevB.63.144303

[13] Smirnov GV, van Bürck V, Franz H, Asthalter T, Leupold O, Schreier E, et al. Nuclear γ resonance time-domain interferometry: Quantum beat and radiative coupling regimes compared in revealing quasielastic scattering. Physical Review B. 2006;**73**:184126. DOI: 10.1103/PhysRevB.73.184126

[14] Saito M, Masuda R, Yoda Y, Seto M. Synchrotron radiation-based quasi-elastic scattering using time-domain interferometry with multi-line gamma rays. Scientific Reports. 2017;**7**:12558. DOI: 10.1038/s41598-017-12216-7

[15] Saito M, Seto M, Kitao S, Kobayashi Y, Higashitaniguchi S, Kurokuzu M, et al. Development of time-domain interferometry for the study of glass formers. Journal of Physics: Conference Series. 2010;**217**:

012147-012150. DOI: 10.1088/1742-6596/217/1/012147

[16] Saito M, Kitao S, Kobayashi Y, Kurokuzu M, Yoda Y, Seto M. Slow processes in supercooled *o*-terphenyl: Relaxation and decoupling. Physical Review Letters. 2012;**109**:115705. DOI: 10.1103/PhysRevLett.109.115705

[17] Saito M, Kitao S, Kobayashi Y, Kurokuzu M, Yoda Y, Seto M. Slow dynamics of supercooled liquid revealed by Rayleigh scattering of Mössbauer radiation method in time domain. Hyperfine Interactions. 2014;**226**:629-636. DOI: 10.1007/s10751-014-1008-9

[18] Saito M, Kobayashi Y, Masuda R, Kurokuzu M, Kitao S, Yoda Y, et al. Slow dynamics in glycerol: Collective de Gennes narrowing and independent angstrom motion. Hyperfine Interactions. 2016;**237**:22. DOI: 10.1007/s10751-016-1243-3

[19] Yamaguchi T, Saito M, Yoshida K, Yamaguchi T, Yoda Y, Seto M. Structural relaxation and viscoelasticity of a higher alcohol with mesoscopic structure. Journal of Physical Chemistry Letters. 2018;**9**:298-301. DOI: 10.1021/acs.jpclett.7b02907

[20] Kanaya T, Inoue R, Saito M, Seto M, Yoda Y. Relaxation transition in glass-forming polybutadiene as revealed by nuclear resonance X-ray scattering. The Journal of Chemical Physics. 2014;**140**: 144906. DOI: 10.1063/1.4869541

[21] Saito M, Mashita R, Masuda R, Kishimoto H, Yoda Y, Seto M. Effect of silica nanoparticle filler on microscopic polymer α-relaxation dynamics. Hyperfine Interactions. 2017;**238**:99. DOI: 10.1007/s10751-017-1466-y

[22] Saito M, Seto M, Kitao S, Kobayashi Y, Higashitaniguchi S, Kurokuzu M, et al. Development of 151-Eu time-domain interferometry and its application for the study of slow dynamics in ionic liquids. Applied Physics Express. 2009;**2**:026502. DOI: 10.1143/APEX.2.026502

[23] Kaisermayr M, Sepiol B, Thiess H, Vogl G, Alp EE, Sturhahn W. Time-domain interferometry using synchrotron radiation applied to diffusion in ordered alloys. The European Physical Journal B. 2001;**20**: 335-341. DOI: 10.1007/s100510170254

[24] Saito M, Seto M, Kitao S, Kobayashi Y, Kurokuzu M, Yamamoto J, et al. Small and large angle quasi-elastic scattering experiments by using nuclear resonant scattering on typical and amphiphilic liquid crystals. Journal of the Physical Society of Japan. 2012;**81**: 023001. DOI: 10.1143/JPSJ.81.023001

[25] Debenedetti PG, Stillinger FK. Supercooled liquids and the glass transition. Nature. 2001;**410**:259-267. DOI: 10.1038/35065704

[26] Angell CA. Relaxation in glassforming liquids and amorphous solids. The Journal of Applied Physics. 2000;**88**:3113-3157. DOI: 10.1063/1.1286035

[27] Ngai KL. Relaxation and Diffusion in Complex Systems. Berlin: Springer; 2011. 835 p. DOI: 10.1007/978-1-4419-7649-9

[28] Angell CA. Relaxation in liquids, polymers and plastic crystals - strong/fragile patterns and problems. Journal of Non-Crystalline Solids. 1991;**131-133**: 13-31. DOI: 10.1016/0022-3093(91)90266–9

[29] Novikov VN, Sokolov AP. Universality of the dynamic crossover in glass-forming liquids: A "magic" relaxation time. Physical Review E. 2003;**67**:031507. DOI: 10.1103/PhysRevE.67.031507

[30] Johari GP, Goldstein M. Viscous liquids and the glass transition. II. Secondary relaxations in glasses of rigid molecules. The Journal of Chemical

Physics. 1970;**53**:2372-2388. DOI: 10.1063/1.1674335

[31] Hansen C, Stickel F, Berger T, Richert R, Fischer EW. Dynamics of glass-forming liquids. III. Comparing the dielectric α- and β-relaxation of 1-propanol and o-terphenyl. The Journal of Chemical Physics. 1997;**107**: 1086-1093. DOI: 10.1063/1.474456

[32] Fujara F, Geil B, Sillescu H, Fleischer G. Translational and rotational diffusion in supercooled orthoterphenyl close to the glass transition. Zeitschrift für Physik B. 1992;**88**:195-204. DOI: 10.1007/BF01323572

[33] Steffen W, Patkowski A, Gläser H, Meier G, Fischer EW. Depolarized-light-scattering study of orthoterphenyl and comparison with the mode-coupling model. Physical Review E. 1994;**49**: 2992-3002. DOI: 10.1103/PhysRevE.49.2992

[34] Petry W, Bartsch E, Fujara F, Kiebel M, Sillescu H, Farago B. Dynamic anomaly in the glass transition region of orthoterphenyl. Zeitschrift für Physik B. 1991;**83**:175-184. DOI: 10.1007/BF01309415

[35] Rössler E, Warschewske U, Eiermann P, Sokolov AP, Quitmann D. Dynamic anomaly in the glass transition region of orthoterphenyl. Journal of Non-Crystalline Solids. 1994;**172-174**:113-125. DOI: 10.1016/0022-3093(94)90424-3

[36] Johari GP. Localized molecular motions of β-relaxation and its energy landscape. Journal of Non-Crystalline Solids. 2002;**307**:317-325. DOI: 10.1016/S0022-3093(02)01491-6

[37] Williams G, Watts DC. Molecular aspects of multiple dielectric relaxation process in solid polymers. Advances in Polymer Science. 1979;**33**:59-92. DOI: 10.1007/3-540-09456-3_3

[38] Vogel M, Rössler E. On the nature of slow α-process in simple glass formers: A 2H NMR study. The Journal of Physical Chemistry. B. 2000;**104**: 4285-4287. DOI: 10.1021/jp9942466

[39] Langer J. The mysterious glass transition. Physics Today. 2007;**60**:8-9. DOI: 10.1063/1.2711621

[40] Kulik AS, Beckham HW, Schmidt-Rohr K, Radloff D, Pawelzik UP, Boeffel C, et al. Coupling of α and β processes in poly(ethyl methacrylate) investigated by multidimensional NMR. Macromolecules. 1994;**27**:4746-4754. DOI: 10.1021/ma00095a015

[41] Garwe F, Schönhals A, Lockwenz H, Beiner M, Schöter K, Donth E. Influence of cooperative α dynamics on local β relaxation during the development of the dynamic glass transition in poly(*n*-alkyl methacrylate)s. Macromolecules. 1996;**29**:247-253. DOI: 10.1021/ma9506142

[42] Richter D, Frick B, Farago B. Neutron-spin-echo investigation on the dynamics of polybutadiene near the glass transition. Physical Review Letters. 1988;**61**:2465-2468. DOI: 10.1103/PhysRevLett.61.2465

[43] Richter D, Zorn R, Farago B, Frick B, Fetters LJ. Decoupling of time scales of motion in polybutadiene close to the glass transition. Physical Review Letters. 1992;**68**:71-74. DOI: 10.1103/PhysRevLett.68.71

[44] Arbe A, Buchenau U, Willner L, Richter D, Farago B, Colmenero J. Study of the dynamic structure factor in the β relaxation regime of polybutadiene. Physical Review Letters. 1996;**76**: 1872-1875. DOI: 10.1103/PhysRevLett.76.1872

[45] Chong SH. Connections of activated hopping processes with the breakdown of the stokes-Einstein relation and with aspects of dynamical heterogeneities. Physics Review. 2008;**E 78**:041501. DOI: 10.1103/PhysRevE.78.041501

[46] Mai YW, Yu ZZ. Polymer Nanocomposites. Cambridge: Woodhead Publishing; 2006. 608 p

[47] Jancar J, Douglas JF, Starr FW, Kumar SK, Cassagnau P, Lesser AJ, et al. Current issues in research on structure–property relationships in polymer nanocomposites. Polymer. 2010;**51**: 3321-3343. DOI: 10.1016/j.polymer.2010.04.074

[48] Arrighi V, Higgins JS, Burgess AN, Floudas G. Local dynamics of poly (dimethyl siloxane) in the presence of reinforcing filler particles. Polymer. 1998;**39**:6369-6376. DOI: 10.1016/S0032-3861(98)00139-6

[49] Gagliardi S, Arrighi V, Ferguson R, Telling MTF. Restricted dynamics in polymer-filler systems. Physica B. 2001; **301**:110-114. DOI: 10.1016/S0921-4526(01)00520-8

[50] Nusser K, Schneider GJ, Richter D. Microscopic origin of the terminal relaxation time in polymer nanocomposites: An experimental precedent. Soft Matter. 2011;**7**: 7988-7991. DOI: 10.1039/C1SM05555K

[51] Schneider GJ, Nusser K, Willner L, Falus P, Richter D. Dynamics of entangled chains in polymer nanocomposites. Macromolecules. 2011; **44**:5857-5860. DOI: 10.1021/ma200899y

[52] Roh JH, Tyagi M, Hogan TE, Roland CM. Space-dependent dynamics in 1,4-polybutadiene nanocomposite. Macromolecules. 2013;**46**:6667-6669. DOI: 10.1021/ma401597r

[53] Roh JH, Tyagi M, Hogan TE, Roland CM. Effect of binding to carbon black on the dynamics of 1,4-polybutadiene. The Journal of Chemical Physics. 2013;**139**:134905. DOI: 10.1063/1.4822476

[54] Glomann T, Hamm A, Allgaier J, Hubner EG, Radulescu A, Farago B, et al. A microscopic view on the large scale chain dynamics in nanocomposites with attractive interactions. Soft Matter. 2013;**9**:10559-10571. DOI: 10.1039/C3SM51194D

[55] Glomann T, Schneider GJ, Allgaier J, Radulescu A, Lohstroh W, Farago B, et al. Microscopic dynamics of polyethylene glycol chains interacting with silica nanoparticles. Physical Review Letters. 2013;**110**:178001. DOI: 10.1103/PhysRevLett.110.178001

[56] Schneider GJ, Nusser K, Neueder S, Brodeck M, Willner L, Farago B, et al. Anomalous chain diffusion in unentangled model polymer nanocomposites. Soft Matter. 2013;**9**: 4336-4348. DOI: 10.1039/C3SM27886G

[57] Guo H, Bourret G, Lennox RB, Sutton M, Harden JL, Leheny RL. Entanglement-controlled subdiffusion of nanoparticles within concentrated polymer solutions. Physical Review Letters. 2012;**109**:055901. DOI: 10.1103/PhysRevLett.109.055901

[58] Deegan RD, Nagel SR. Dielectric susceptibility measurements of the primary and secondary relaxation in polybutadiene. Physical Review B. 1995; **52**:5653. DOI: 10.1103/PhysRevB.52.5653

[59] Kohn VG, Chumakov AI, Rüffer R. Wave theory of focusing monochromator of synchrotron radiation. Journal of Synchrotron Radiation. 2009;**16**:635-641. DOI: 10.1107/S090904950902319X

[60] Shvyd'ko Y. X-ray echo spectroscopy. Physical Review Letters. 2016;**116**:080801. DOI: 10.1103/PhysRevLett.116.080801

[61] Bencivenga F, Cucini R, Capotondi F, Battistoni A, Mincigrucci R, Giangrisostomi E, et al. Four-wave mixing experiments with extreme ultraviolet transient gratings. Nature. 2015;**520**:205-208. DOI: 10.1038/nature14341

Section 2

X-Ray Powder Diffraction

Chapter 5

Identification and Quantification of Phases in Steels by X Ray Diffraction Using Rietveld Refinement

Adriana da Cunha Rocha and Gabriela Ribeiro Pereira

Abstract

X-ray diffraction has been applied in the investigation of phase formation in steels, operating in industrial environments. In this work, identification and quantification of phases by X-ray diffraction and peak fitting, using the Rietveld method, were employed. In a first scenario, two different types of steels, subjected to abrasive surface cleaning, suffered contamination from the blasting operation that compromise between 10 and 36% of the blasted surface, as revealed by the quantitative phase analysis. Such high values of particulates can jeopardize the corrosion protection offered by posterior coating application. In a second scenario, duplex steels (DS) subjected to aggressive environments and high temperatures of service went through phase transformation that formed amounts up to 3.5% of a deleterious phase, known as sigma phase. This phase compromises the steel mechanical resistance and corrosion protection, and its quantification is crucial for the assurance of the material integrity. The quantitative phase analysis (QPA) by X-ray diffraction provided the diagnosis of forthcoming problems related to the presence of such phases in the investigated steels, allowing the optimization of techniques and the choice of correct process parameters.

Keywords: X-ray diffraction, steels, Rietveld refinement, phase quantification

1. Introduction

Steel is the most common material used on earth. Applications vary from simple cutlery to spacecraft parts and are so vast; one finds even hard to list it all. This is mainly due to the versatility found in this type of iron and carbon alloy, in terms of physical, mechanical, and chemical properties. Also, when compared to other types of materials, steels are economically affordable. Therefore, steel has been studied for many decades and will continue to be so in the forthcoming years. Industrial plants have most of their equipment made of steel. Applications involving the oil and gas industry are very demanding in terms of optimizing the use of these steels for high performance in constant aggressive environments. In this case, the ultimate need is for steels that can resist both heavy loads and aggressive corrosive environments.

Some new classes of steels, such as the duplex steels, are of very much of interest nowadays, because of their good compromise between mechanical resistance and corrosion protection [1, 2]. As any other metallic material though, they usually need thermal-mechanical processing in order to be adequate to the different uses. Thermal-activated processes may lead to the creation of new phases in the steel—some intentionally promoted, but some not. The knowledge of phase transformations in steels is mandatory to forecast the properties the material will acquire after such transformations [3]. Because steel has a long of range periodic atomic structure, with well-defined crystallographic aspects [4], X-ray diffraction [5] is one of the most important analytical techniques to identify those structures, in order to understand steel properties.

Lately, the identification and quantification of phases have been upgraded by many methods of peak refinement. These methods provide a good calculation of crystallographic parameters, enabling precise measurements to be performed in different materials. Among those methods, the Rietveld refinement [6] has been gaining space among crystallographers due to its analytical capabilities. A general overview of the Rietveld profile fitting and quantitative phase analysis is provided in the following sections. Then, two specific applications of X-ray diffraction for steel phase analysis are described. The first case refers to the quantification of contaminants on steel substrates after jet impingement, aiming corrosion resistance by organic coatings. The second case is related to the phase transformations occurring in a type of steel used in oil and gas applications, when this material is subjected to high temperatures due to welding procedures or operation in service. In both situations, peak refinement is made, for the calculation of crystallographic parameters and for quantitative phase calculations.

2. Peak refinement and quantitative phase analysis: the Rietveld method

X-ray profile fitting provides important crystallographic information from the analyzed material. There are several different techniques nowadays, but one of them, known as the Rietveld refinement method, has many advantages over the others. In this method, first presented by Hugo Rietveld to refine nuclear and magnetic structures [6] and lately developed by many scientists [7], least-squares refinements are carried out until the best fit is obtained between the entire observed powder diffraction pattern and the full calculated pattern. The quantity minimized in the least-squares refinement is the residual S_y:

$$S_y = \sum_i w_i \left(y_i - y_{ci} \right)^2 \quad (1)$$

where y_i = observed intensity at the i-th step; y_{ci} = calculated intensity at the i-th step; $w_i = 1/y_i$.

The equation model applied for the method (Eq. 2) considers the following parameters:

- The Bragg reflections contributing to a specific intensity y_i at every specific i point in the whole pattern
- A scale factor s
- The Miller indices, h, k, l, for a Bragg reflection, represented by K
- The Lorentz polarization and multiplication factors L_K
- The reflection profile function ∅

- The preferred orientation function P_k
- The absorption factor A
- The structure factor modulus for the kth Bragg reflection $|F_K|$
- The background intensity at the i-th step, y_{bi}

$$y_{ci} = s\sum_{K} L_K|F_K|^2\, Ø\,(2\Theta_i - 2\Theta_k)\, P_K A + y_{bi} \tag{2}$$

Quantitative phase analysis using the Rietveld method [8] employs the relative weight fraction W of each phase p in a mixture of n phases calculated according to the equation:

$$W_p = \left(s_p\, ZMV\right) / \sum_{i=1}^{n} s_i\, (ZMV)_i \tag{3}$$

where s is the Rietveld scale factor, Z is number of formula units per unit cell, M is the mass of the formula unit (atomic mass unit), and V is the unit cell volume ($Å^3$).

The following sections will present two specific cases of the utilization of the Rietveld refinement method with further quantitative phase analysis (QPA). Those practical cases demonstrate how this methodology was applied for the analysis of steel parts, addressing the presence of unwanted phases and phase unbalance due to thermal treatments performed in specific steels.

3. Case 1: abrasive blasting in steel surfaces—addressing contamination by X-ray diffraction quantitative analysis

Duplex and super duplex steels (DS and SDS, respectively) have been widely used in oil and gas industries because of their advantages over other steel types in terms of mechanical properties and corrosion resistance [1, 2]. The harsh environments where those steels are in service require protection from degradation that can be found in organic coatings [9, 10].

The coating performance is highly dependent on the surface pretreatment and the application procedures [11, 12]. Those must be in accordance with standard documents [13, 14], which include procedures for blast cleaning. Blasting processes though might affect the coating adhesion and corrosion rate, depending on the degree of contamination from the abrasive particulate material used, as those particulates can promote local pH changes and/or galvanic effects [15]. The common abrasives employed for surface treatment of steels are aluminum oxide and martensitic steel abrasives due to their high values of hardness. Pulverization of the grits, however, can lead to undesired particulate depositions over the steel surfaces (**Figure 1**), which induce local alkalization, decreasing the protection. Because of all these factors, substrate contamination needs to be engaged in an efficient fashion, to avoid damages on the performance of the whole system.

Determination of the inclusion or second-phase constituent, by metallographic analysis [16], can be used to account for such contamination. However, the technique can be quite time-consuming. Quantitative phase analysis by X-ray diffraction though can be used for such task [17–22]. The Rietveld method can provide very accurate estimative of the relative and/or absolute amount of the component phases [22–25] and has advantages over traditional internal-standard-based

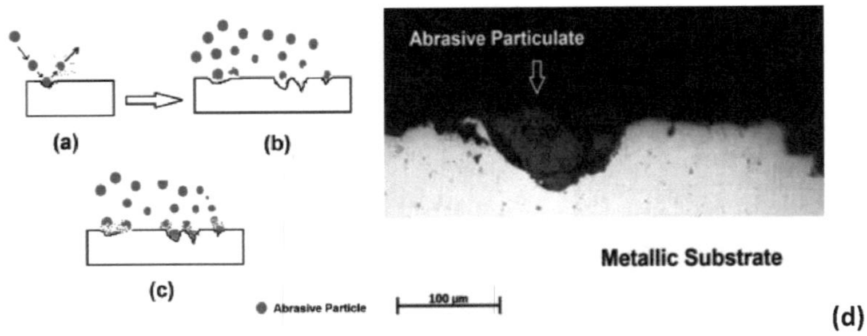

Figure 1.
(a, b) Abrasive particles hitting a metal substrate surface and (c) abrasive fragments deposited over the surface. (d) A real micrographs of a particulate allocated in the valley created by the particle impact in the surface.

techniques. Surface roughness effects can also be considered and compensated by correction functions, which makes the Rietveld method more interesting to this type of process.

3.1 Surface roughness corrections

In Rietveld analysis of X-ray powder diffraction patterns, the effect of surface roughness (SR) of absorbing polycrystalline samples can be a source of systematic errors [26–30]. The SR effect can reduce the intensity of low-angle reflections and lead to anomalous low values of refined atomic displacement parameters. Depending on the degree of SR, the isotropic atom displacement can lead to negative values, which have no physical meaning. To correct such effects, a SR Suortti Model [31] has been used to guarantee a higher flexibility in terms of angular ranges.

3.2 Experimental parameters

ASTM A516 G60 carbon steel (CS) and UNS 32760 super duplex steel (SDS) samples were used as metallic substrates subjected to the blasting process. The abrasives used encompassed two types of aluminum oxide particulate (sintered bauxite (SB) and demagnetized alumina (DA)). A D8 Discover Bruker AXS was the equipment used for data acquisition. The diffraction parameters are listed as following:

- Radiation: Co Kα (λ = 1789 Å).
- Current and voltage: 40 mA 35 kV.
- Primary optics: Co Göbel Mirror, two slits of 1 mm and 6 mm and a soller slit with 2 cm x 1 cm aperture.
- Secondary optics: Kβ filter, 8 mm slit, axial soller slit with divergence of 2.5°.
- Detector: point scanning detector—PSD type.
- 2θ range = 10° to 110°.
- Step-size: 0.001°.
- Scanning velocity was 0.5 s/step.

Rietveld analysis was carried out using Diffrac PlusTOPAS (ver 4.2) software [32, 33].

3.3 X-ray analysis results

Diffraction patterns were obtained for both substrate bulks, prior to the blasting process, to work as a reference pattern when measuring the degree of contamination of the samples subsequently analyzed. In the blasted surfaces, α-Fe (ferrite) [34] was observed in CS substrate, while α-Fe and γ-Fe [35] (ferrite and austenite, respectively) were present in the SDS substrate.

The commercial SB abrasive showed a predominance of phase alpha alumina (α-Al_2O_3) [36] which was verified in the SB abrasive, while the DA abrasive presented a majority of kappa alumina (κ-Al_2O_3) [37]. **Figure 2** presents the diffraction patterns for the carbon steel substrate before and after abrasive blasting (a) and for the super duplex steel before and after blasting (b), respectively.

Figure 3 shows the detailed refined scan for the carbon steel substrate blasted with κ-Al_2O_3 from the DA abrasive and α-Al_2O_3 originated from the SB abrasive. In the same manner, **Figure 4** presents the result of the refined scan from the SDS substrate blasted with κ-Al_2O_3 from the DA abrasive and α-Al_2O_3 originated from the SB abrasive.

3.3.1 Fitting parameters

The structure refinement functions and parameters are listed as following:

- Chebyshev polynomial of fourth degree [38] and Topas 1/x background function and (background fitting intensities, y_{ib})
- Preferred orientation (PO) March-Dollase model [39–41] for calculating the preferred crystal orientations of α-Fe and γ-Fe phases (this is mandatory especially for processed steel products like ingots, sheets, and pipe sections)
- PO spherical harmonics [42] model of order 6 for the alumina phase

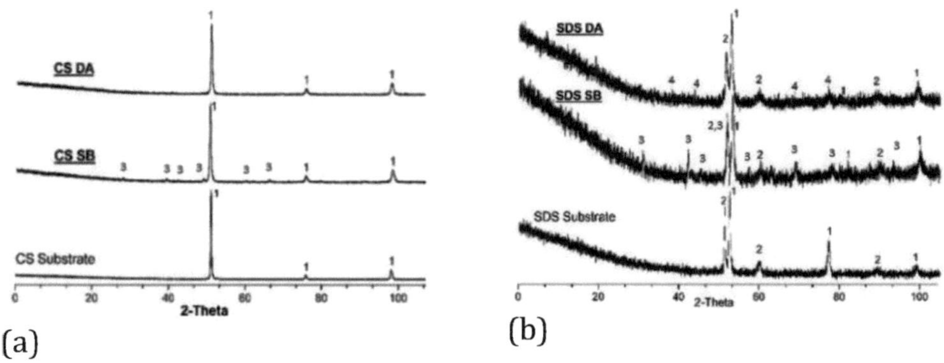

Number	1	2	3	4
Phase	Fe-α	Fe-γ	α-Al_2O_3	κ-Al_2O_3

Figure 2.
(a) CS substrate after DA and SB blasting and (b) SDS substrate after DA and SB blasting. When blasting is performed with Al_2O_3 abrasives, one can see contamination by the new peaks introduced to the scans.

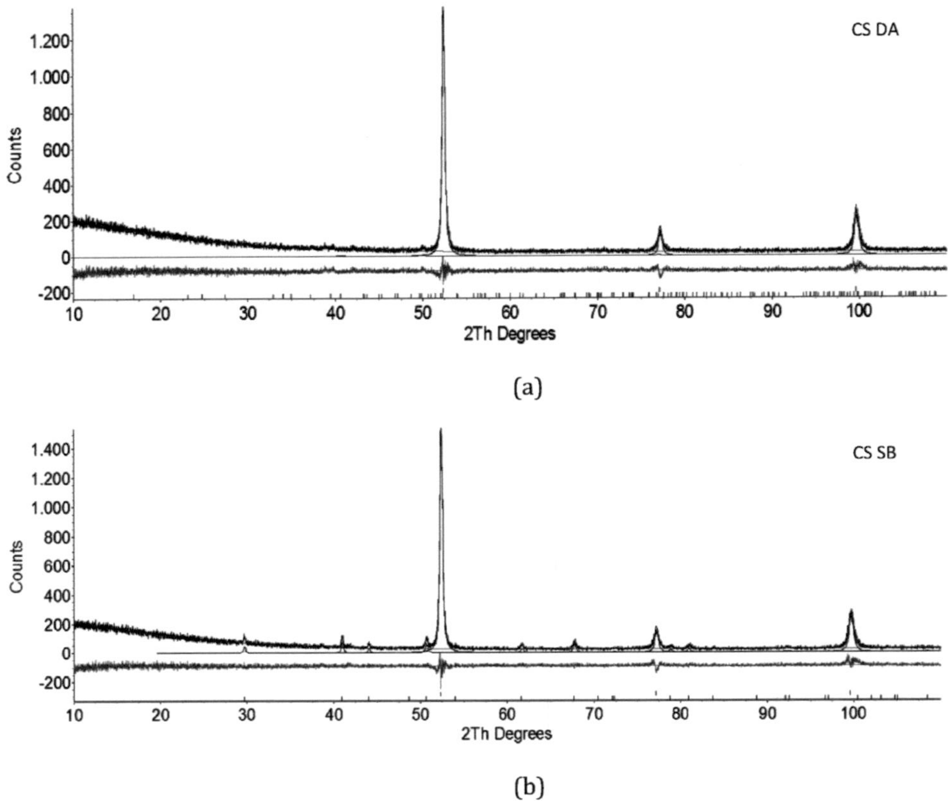

Figure 3.
Carbon steel substrate blasted with (a) DA and (b) SB abrasives. Observed data are indicated by thicker lines and calculated data by a solid thinner line. The gray lower curve presents the difference (residue) between the observed and calculated powder diffraction patterns.

Zero error (2θ) sample displacement, absorption (1/cm), and lattice parameters of the phases were not fixed to provide the best calculated fitting.

3.3.2 Fitting criteria

Fitting criteria is a way to analyze the accuracy and precision of fitting. Based on the R-weighted pattern (R_{wp}) and the R-expected pattern (R_e), it is possible to calculate the "goodness of fit," or simply *GOF*, to address the calculated values. Eqs. 4 and 5 present the variables used for the calculations for the R-values, which are then used to calculate the GOF [43–45]:

$$R_{wp} = \left[\left(\sum w_i (y_i(obs) - y_i(calc))_2\right) / \left(\sum w_i (y_i(obs)_2)\right)\right] 1/2 \quad (4)$$

$$R_{exp} = \left[(N - P) / \left(\sum w_i (y_i(obs)2)\right)\right] 1/2 \quad (5)$$

$$GOF = [(S_y)/(N - P)] 1/2 = R_{wp}/R_{exp} \quad (6)$$

where y_i = intensity at the ith step; w_i = weighting factor; N = number of observations; P = number of parameters; obs = observed and calc = calculated.

Table 1 presents the GOF values for each calculation. The calculated values lied between 1 and 1.5, which is an indication of a satisfactory fitting. Numbers greater than 1.5 are usually seen as an inadequate model or false minimum, whereas those

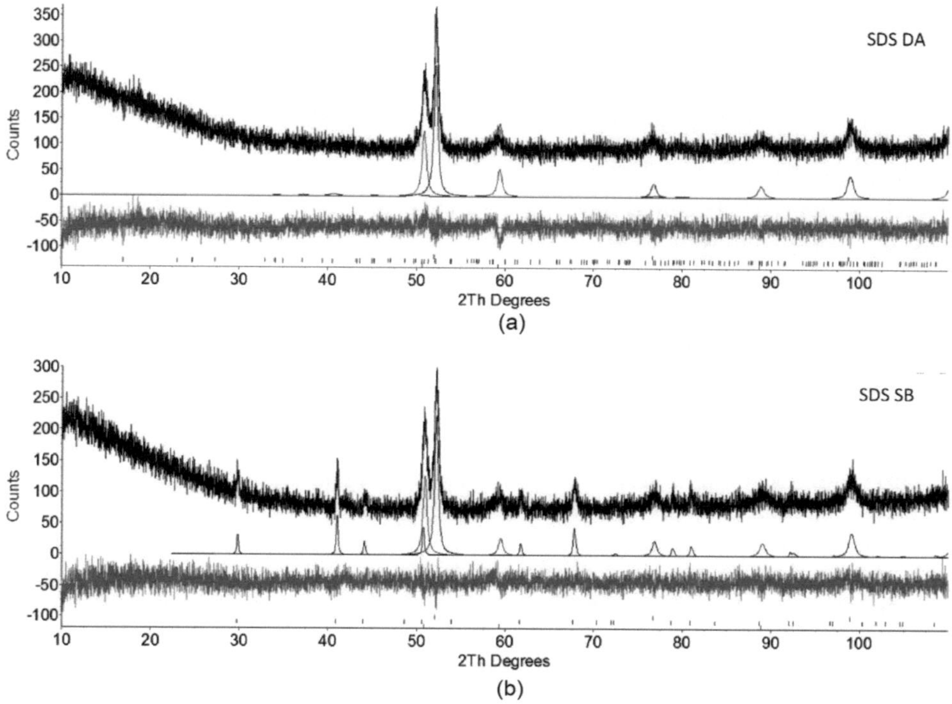

Figure 4.
Super duplex steel substrate blasted with (a) DA and (b) SB abrasives.

Abrasive	CS substrate	SDS substrate
SB	1.14 ± 0.008	1.05 ± 0.007
DA	1.13 ± 0.010	1.07 ± 0.010
MCS	1.09 ± 0.012	1.06 ± 0.010
MSS	1.09 ± 0.011	1.06 ± 0.011

Table 1.
Fitting criteria for Rietveld calculations: calculated average goodness of fitness for the set of four samples of each substrate per abrasive.

	% α-Fe	% γ-Fe	%Al_2O_3
SB abrasive			
Super duplex	38.79 ± 1.84	25.01 ± 2.13	36.20 ± 2.92
Carbon steel	79.79 ± 2.37	*	20.21 ± 2.37
DA abrasive			
Super duplex	47.31 ± 2.21	36.92 ± 1.16	15.77 ± 2.52
Carbon steel	89.56 ± 0.59	*	10.45 ± 0.59

Table 2.
Quantitative phase calculations results (calculated average and standard deviation for a set of four blasted samples).

lower than 1.0 show a model that contain more parameters than can be justified by the quality of the data, as insufficient counting time for processing or high influence of background, for example.

Table 2 presents the quantitative phase analysis results for abrasive contamination in both CS and DSS substrates. 36.20% of the SDS and 20.21% of the carbon steel blasted area were contaminated by SB particles. When analyzing the DA abrasive, 15.77% of the SDS area was contaminated, while 10.45% of the CS substrate depicted particle contamination. The higher percentage of contamination on the SDS substrate can be related with its high values of hardness. The consequences of such higher particle contamination, for the performance of anticorrosive organic coatings, can be found in a subsequent work [46].

4. Case 2: ferrite/austenite (α/γ) ratio in duplex steels and the occurrence of sigma phase: quantification of unbalanced phase formation and precipitation due to thermal treatments on the steel

Super duplex stainless steel (SDSS) is a class of steels that retain two equal balanced main phases within their microstructure, BCC α-Fe (ferrite) and FCC γ-Fe (austenite). In that manner, this material can combine good mechanical properties with high corrosion resistance. However, when subjected to welding or to high-temperature applications, thermal-activated diffusion mechanisms promote the precipitation of some deleterious phases in the SDSS matrix in addition to creating an unbalanced volume of ferrite and austenite. The unequal proportions of ferrite/austenite and the occurrence of phases such as sigma phase (also known as σ phase) can highly compromise the ability of these steels to support loads and to avoid corrosion, leading to higher rates of degradation. Therefore, it is mandatory that investigations on thermal cycles are carried on determining the critical time/temperature values that lead to this kind of phase unbalance.

Previous studies in different classes of duplex steels [47] have identified the temperature range of 300–1000°C as a critical range for phase transformations. Therefore, a series of heat treatments, involving different temperature ranges and time intervals, were performed in a UNS S32750 to study the phase formation in this specific class of duplex steel and to determine the amounts of ferrite, austenite, and sigma phase formed after each treatment. For this specific calculation, X-ray diffraction was displayed as a crucial tool for precise phase quantification in a specific volume of material. After all the samples were scanned, phase amounts were calculated using quantitative phase analysis by Rietveld refinement. These calculations lead to further experimental investigations using nondestructive evaluation techniques [48].

4.1 Heat treatments for different amounts of phase formation

Samples were cut as 70 mm × 40 mm × 6 mm steel plates. All samples were submitted to a preliminary solution heat treatment in order to obtain a balance of approximately 50% of α and γ phases. Then, aging treatments were performed to create the α/γ unbalance and the precipitation of sigma phase. **Figure 5** shows a schematic of the heat treatment steps.

The solution heat treatment was conducted as follows:

1. Three samples remained in the as-received condition, i.e., without any heat treatment for further comparison with the heat-treated samples.

2. The remaining samples were subjected to a solubilization treatment, which consists of heating up to 1220°C for 1 h, followed by water quenching.

3. Three of the solubilized samples did not receive any additional aging heat treatment and remained in the solubilized condition.

4. Then, a group of 14 samples received an additional aging heat treatment to introduced different fractions of sigma phase. The aging heat treatment was conducted at 1000°C for different time intervals, followed by water quenching.

5. Finally, seven samples were heat treated at 1320 and at 1350°C for different intervals, in order to have high amounts of delta phase but no sigma phase at all.

4.2 X-ray analysis results

Phase volumetric fractions were measured in nine different regions of each sample, as depicted in **Figure 6**. Diffraction parameters used were the same presented in item 3.2 from this chapter.

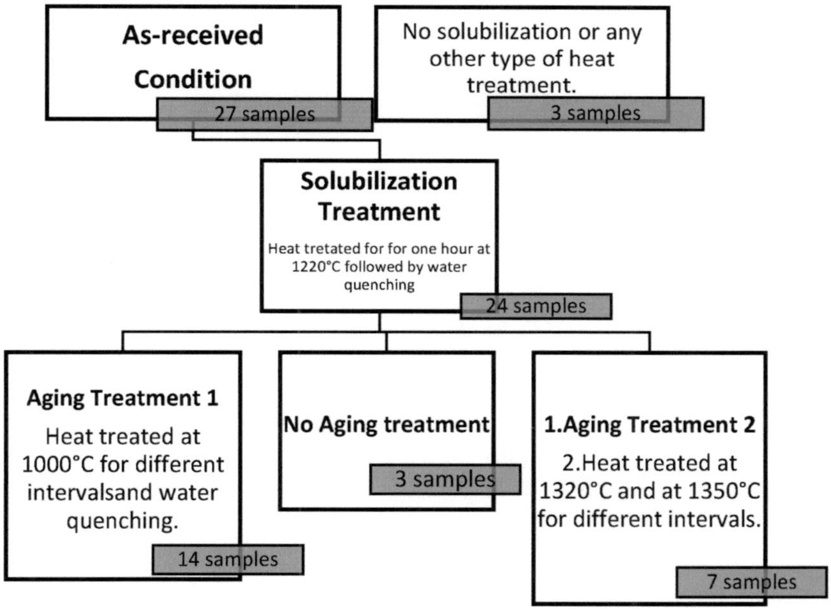

Figure 5.
Schematics of heat treatments performed in the SDSS samples.

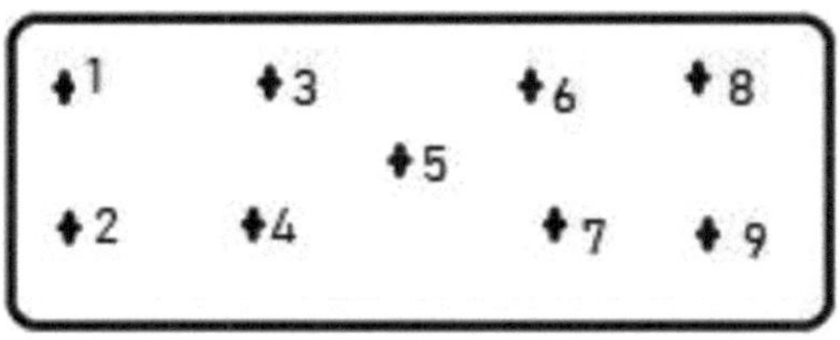

Figure 6.
Schematics of a sample with its nine analyzed points.

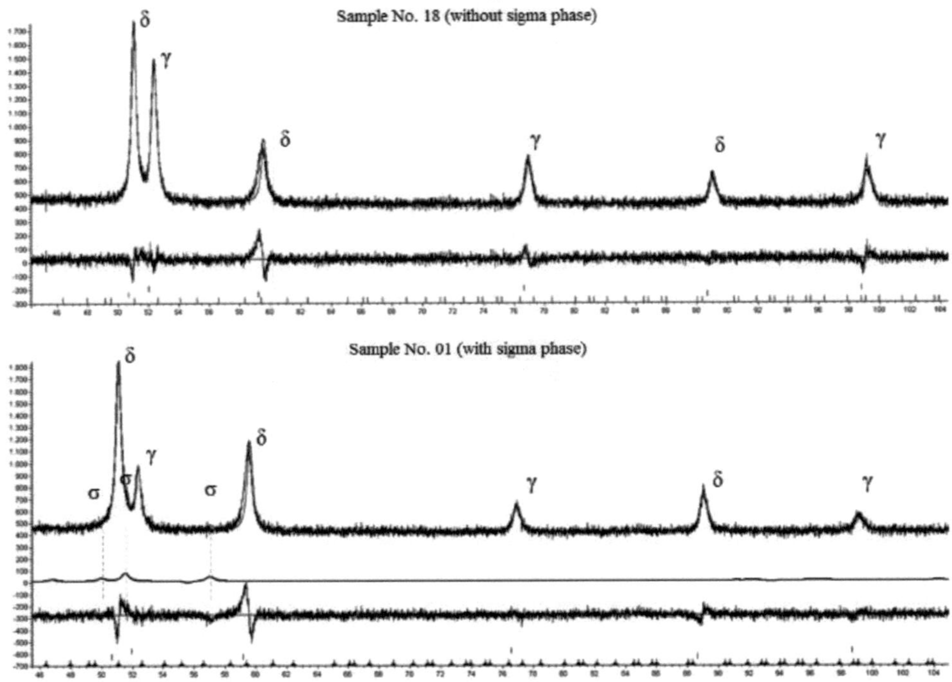

Figure 7.
XRD spectrum for two different conditions. Sample number 18 without σ phase and sample number 01 with 3.4% of σ phase.

Samples	Temperature (°C)	Time (min)	γ phase (%)	α phase (%)	σ phase (%)
01	1000	60	64.0 ± 2.3	32.5 ± 2.7	3.4 ± 1.0
02	1000	45	49.3 ± 3.0	47.5 ± 3.5	3.1 ± 0.9
03	1000	22	64.3 ± 3.9	32.6 ± 4.0	3.0 ± 0.6
04	1000	45	62.4 ± 4.3	34.8 ± 4.1	2.7 ± 0.7
05	1000	25	52.2 ± 12.1	45.1 ± 11.5	2.6 ± 1.2
06	1000	25	65.1 ± 9.8	31.7 ± 7.8	2.4 ± 1.1
07	1000	5	68.1 ± 7.9	29.6 ± 8.2	2.2 ± 0.7
08	1000	60	61.1 ± 5.0	36.6 ± 4.9	2.1 ± 1.9
09	1000	20	64.4 ± 4.5	33.4 ± 4.5	2.1 ± 0.2
10	1000	20	56.7 ± 6.5	41.2 ± 6.9	2.0 ± 0.7
11	1000	1	57.9 ± 5.5	40.4 ± 5.3	1.6 ± 0.6
12	1000	1	59.3 ± 7.1	39.0 ± 7.1	1.6 ± 0.2
13	1000	6	68.5 ± 3.6	29.9 ± 3.6	1.5 ± 0.4
14	1000	10	61.6 ± 5.4	37.0 ± 5.3	1.2 ± 0.4
15	As received		47.7 ± 2.0	52.2 ± 2.0	0.0
16	As received		44.2 ± 4.9	55.7 ± 4.9	0.0
17	As received		47.1 ± 1.6	52.8 ± 1.6	0.0
18	1220	60	50.2 ± 7.8	49.7 ± 7.7	0.0
19	1220	60	56.8 ± 5.1	43.1 ± 5.1	0.0
20	1220	60	54.3 ± 5.7	45.7 ± 5.7	0.0

Samples	Temperature (°C)	Time (min)	γ phase (%)	α phase (%)	σ phase (%)
21	1320	60	38.8 ± 3.3	61.1 ± 2.9	0.0
22	1320	60	28.3 ± 5.1	71.6 ± 5.0	0.0
23	1320	120	44.8 ± 3.0	55.1 ± 3.0	0.0
24	1320	60	36.2 ± 7.4	63.7 ± 7.4	0.0
25	1320	240	41.7 ± 6.9	58.2 ± 6.9	0.0
26	1350	60	34.4 ± 6.4	65.7 ± 6.3	0.0
27	1350	60	40.5 ± 8.7	59.4 ± 8.6	0.0

Table 3.
Phase volume contents according to heat treatment temperatures and time intervals.

4.2.1 Fitting parameters

The structure refinement used the fifth-degree Chebyshev polynomial [38] to fit the background intensities, y_{ib}, (according to Eq. 2), as well as the 1/x background function, from Topas 4.2. α-Fe and γ-Fe and sigma phases were fitted to the preferred orientation March-Dollase model [39–41].

4.2.2 Fitting criteria

The fitting criteria followed the same methodology applied in Case 1, using Eqs. (4)–(6). For every sample, the GOF was within the range of 1.0–1.5.

4.2.3 Phase calculations

Figure 7 depicts two diffractograms—one from a sample containing only ferrite and austenite and another containing both phases and sigma. QPA (using Rietveld refinement) was carried on each one of the nine described points for each sample, generating similar scans to the ones presented in **Figure 7**. Each scan was then carefully analyzed and adjusted accordingly to the chosen fitting parameters to assure a GOF between 1.0 and 1.5, i.e., the best fit possible.

The values obtained for each point were then summed and averaged and the standard deviation calculated for each sample average. **Table 3** presents those calculated values.

5. Conclusions

X-ray diffraction has demonstrated to be an effective tool for phase analysis in metallic materials, especially in steels. Because this type of material is the most used material on earth nowadays, due to its versatility in terms of physical, mechanical, and chemical properties, knowledge of the phase transformations that might occur during service and processing is ultimate.

Steel surfaces subjected to abrasive surface cleaning, which suffered contamination from the blasting operation, and duplex steels subjected to aggressive environments and high temperatures of service, which experienced phase transformation, were analyzed by X-ray diffraction using peak refinement, by the Rietveld method.

The refinement method demonstrated that phase identification and quantification enabled the diagnosis of forthcoming problems related to the presence of such phases in the investigated steels, allowing the optimization of techniques and the choice of correct process parameters.

Acknowledgements

The authors would like to acknowledge the staff and professors from the Laboratorio de Ensaios Nao Destrutivos Corrosao e Soldagem (LNDC, Rio de Janeiro, Brazil) that were fundamental for the realization of this work, as well as the Department of Materials Sciences and Metallurgy (DMM) of the Federal University of Rio de Janeiro (UFRJ).

Conflict of interest

The authors declare no conflict of interest.

Thanks

The authors would like to thank professor Isabel Margarit-Mattos for the opportunity of developing this investigation on the blasted steel surfaces from her research on organic coatings.

Author details

Adriana da Cunha Rocha* and Gabriela Ribeiro Pereira
Federal University of Rio de Janeiro (UFRJ), Rio de Janeiro, Brazil

*Address all correspondence to: adrirocha@metalmat.ufrj

References

[1] Armas IA, Moreuil SD. Duplex Stainless Steels. New Jersey: Wiley; 2009

[2] Bernhardsson S, Oredsson J, Martenson C. Duplex Stainless Steels. Metals Park, Ohio, USA: ASM International; 1983

[3] Porter DA, Easterling KE. Phase Transformations in Metals and Alloys. 2nd ed. United States: Springer; 1992

[4] ASM International Handbook Committee. ASM Handbook, Vol. 1: Properties and Selection: Irons, Steels, and High-Performance Alloys. Novelty, OH, USA: ASM International; 2009

[5] Cullity BD. Elements of X-Ray Diffraction. 2nd ed. Boston: Addison-Wesley Publishing Company; 1978

[6] Rietveld HM. A profile refinement method for nuclear and magnetic structures. Journal of Applied Crystallography. 1969;**2**(2): 65-71

[7] Young R. The Rietveld Method. New York: Oxford University Press; 1995

[8] Hill RJ, Howard CJ. Quantitative phase analysis from neutron powder diffraction data using the Rietveld method. Journal of Applied Crystallography. 1987;**20**:467-474

[9] Schweitzer PA. Corrosion Engineering Handbook. 2nd ed. Florida: CRC Press; 2007

[10] Satas D, Tracton AA. Coatings Technology Handbook. 3rd ed. Florida: CRC Press; 2005

[11] Momber A. Blasting Cleaning Technology. Berlin: Springer; 2008

[12] Momber A, Wong YC. Overblasting effects on surface properties of low-carbon steel. Journal of Coating Technology and Research. 2005;**2**(6): 453-461

[13] ISO 4628-2. Paints and Varnishes, Evaluation of Degradation of Coatings–Designation of Quantity and Size of Defects and of Intensity of Uniform Changes in Appearance–Assessment of Degree of Blistering. Switzerland: International Organization for Standardization; 2003

[14] ISO 8501-1. Preparation of Steel Substrates before Application of Paints and Related Products. Visual Assessment of Surface Cleanliness. Rust Grades and Preparation Grades of Uncoated Steel Substrates and of Steel Substrates After Overall Removal of Previous Coatings. Switzerland: International Organization for Standardization; 2007

[15] Morcillo M, Bastidas M, Simancas J, Galvan JC. The effect of the abrasive work mix on paint performance over blasted steel. Anti-Corrosion Methods and Materials. 1989;**36**(5):4-8

[16] ASTM E1245–03. Practice for Determining the Inclusion or Second-Phase Constituent Content of Metals by Automatic Image Analysis. USA: American Society for Testing and Materials, ASTM International; 2003

[17] Zevin LS, Kimmel G. Quantitative X-Ray Diffractometry. Berlin: Springer; 1995

[18] Klug HP, Alexander LE. X-Ray Diffraction Procedures for Polycrystalline and Amorphous Materials. 2nd ed. New Jersey: John Wiley & Sons; 1974

[19] Hubbard CR, Snyder RL. RIR—Measurement and use in quantitative XRD. Powder Diffraction. 1988;**3**(2): 74-77

[20] Chung FH. Quantitative interpretation of X-ray diffraction patterns of mixtures. I. Matrix-flushing method for quantitative multicomponent analysis. Journal of Applied Crystallography. 1974;7:519-525

[21] Bish DL, Howard SA. Quantitative phase analysis using the Rietveld method. Journal of Applied Crystallography. 1988;**21**:86-91

[22] Bish DL, Chipera SJ. Problems and solutions in quantitative analysis of complex mixtures by X-ray powder diffraction. Advances in X-Ray Analysis. 1987;**31**:295-308

[23] Bish DL, Chipera SJ. Accuracy in quantitative X-ray powder diffraction analyses. Advances in X-Ray Analysis. 1995;**38**:47-57

[24] Chipera SJ, Bish DL. Multireflection RIR and intensity normalizations for quantitative analyses: Applications to feldspars and zeolites. Powder Diffraction. 1995;**10**(1):47-55

[25] Snyder RL, Bish DL. Quantitative analysis—Modern powder diffraction. In: Bish DL, Post JE, editors. Reviews in Mineralogy. Vol. 20. Denver: Mineralogical Society of America; 1989. pp. 101-144

[26] Sidey V. A simplified correction function for the effect of surface roughness in X-ray powder diffraction. Journal of Applied Crystallography. 2004;**37**:1013-1014

[27] Paiva-Santos CO, Marques RFC, Franco MK, Silva MT, Jafelicci-Jr M. Surface Roughness Effect on the B's Values in the Rietveld Refinement. 2002. Available from: http://labcacc.iq.unesp.br/publicacoes/sr/sreffect.htm

[28] Sparks CJ, Kumar R, Specht ED, Zschack P, Ice GE, Shiraishi T, et al. Effect of powder sample granularity on fluorescent intensity and on thermal parameters in X-ray diffraction Rietveld analysis. Advances in X-Ray Analysis. 1992;**35**:57-62

[29] Pitschke W, Hermann H, Mattern N. The influence of surface roughness on diffracted X-ray intensities in Bragg-Brentano geometry and its effect on the structure determination by means of Rietveld analysis. Powder Diffraction. 1993;**8**(2):74-83

[30] Pitschke W, Mattern N, Hermann H. Incorporation of microabsorption corrections into Rietveld analysis. Powder Diffraction. 1993;**8**(4):223-228

[31] Suortti P. Effects of porosity and surface roughness on the X-ray intensity reflected from a powder specimen. Journal of Applied Crystallography. 1972;**5**:325-331

[32] Bruker AXS. Topas: General Profile and Structure Analysis Software for Powder Diffraction Data. Karlsruhe: GmbH; 2003

[33] Cheary RW, Coelho A. A fundamental parameters approach to X-ray line-profile fitting. Journal of Applied Crystallography. 1992;**25**(2):109-121

[34] Gorton AT, Bitsianes G, Joseph TL. Thermal expansion coefficients for iron and its oxides from X-ray diffraction measurements at elevated temperatures. Transactions of the Metallurgical Society of AIME. 1965;**233**(8):1519-1525

[35] Ridley N, Stuart H. Partial molar volumes from high-temperature lattice parameters of iron–carbon austenites metal. Science Journal. 1970;**4**:219-222

[36] Morris MC et al. Standard X-ray diffraction powder patterns. National Bureau of Standards (U.S.), Circ 539. 1960;**9**:3

[37] Olivier B, Retoux R, Lacorre P, Massiot D, Ferey G. Crystal structure of κ-alumina: An X-ray powder diffraction, TEM and NMR study. Journal of Materials Chemistry. 1997;7:1049

[38] Boyd JP. Chebyshev and Fourier Spectral Methods. 2nd ed. New York: Dover Publications; 2001

[39] March A. Mathematische Theorie der Regelung nach der Korngestalt bei affiner Deformation. Zeitschrift fuer Kristallographie. 1932;**81**:285-297

[40] Dollase WA. Correction of intensities for preferred orientation in powder diffractometry: Application of the March model. Journal of Applied Crystallography. 1986;**19**:267-272

[41] Zolotoyabko E. Determination of the degree of preferred orientation within the March-Dollase approach. Journal of Applied Crystallography. 2009;**42**:513-518

[42] Solomentsev ED. In: Hazewinkel M, editor. Encyclopedia of Mathematics. Berlin: Springer; 2001

[43] E. Prince E. In: Wilson AJC, editor. Structure and Statistics in Crystallography. New York; 1985. pp. 95-108

[44] Young RA, Wiles DB. Profile shape functions in Rietveld refinements. Journal of Applied Crystallography. 1982;**15**:430-438

[45] Young RA, Prince E, Sparks RA. Suggested guidelines for the publication of Rietveld analyses and pattern decomposition studies. Journal of Applied Crystallography. 1982;**15**: 357-359

[46] Nascimento AV. Aspects of surface treatments of super duplex stainless steels for application of anticorrosive coatings [Thesis]. Rio de Janeiro, Brazil: UFRJ; 2009

[47] Yamashita M, Matsumoto S, Hironata N. Corrosion resistance and properties of duplex stainless steels. In: International Conference &Expo Duplex Stainless Steels'07; Italy, June. 2007

[48] Camerini C, Sacramento R, Areiza MC, Rocha A, Santos R, Rebello JM, et al. Eddy current techniques for super duplex stainless steel characterization. Journal of Magnetism and Magnetic Materials. 2015;**388**:96-100

Chapter 6

In Situ Titanium Composites: XRD Study of Secondary Phases Tied to the Processing Conditions and Starting Materials

Eva María Pérez-Soriano, Cristina M. Arévalo-Mora and Isabel Montealegre-Meléndez

Abstract

Nowadays, the development of high specific modulus materials involves studies of new materials and novel manufacturing routes. From the point of view of composite materials, titanium composites (TMCs) have been long studied for their interesting properties, as a result of the conjunction of low-density and high mechanical properties, as well as corrosion resistance. Among various processing techniques, the in situ reinforced method shows many advantages above the rest. The reactions between matrix and reinforcement drive up the final properties of TMCs. Varying the processing conditions, in addition to reinforcement type and content, significant variations are expected in TMCs' behaviour. In this regard, the present study draws on previous author works. The specimens studied were manufactured by hot consolidation processes, inductive hot pressing (iHP) and direct hot pressing (DHP), at different operational parameters and compositions. X-ray powder diffraction (XRD) investigations tied formations of secondary phases to substantive changes in TMC behaviour under the influence of the fabrication parameters.

Keywords: XRD analysis, titanium composites, secondary phases, in situ reaction, matrix strengthening

1. Introduction

Over the last decades, investigations about composite materials have made great advances in understanding the importance of the starting materials and the manufacturing process, for the development of novel materials with outstanding properties [1]. In this regard, in the field of metal matrix composites, research studies have been conducted to achieve optimal bounding matrix reinforcement, improving the strength of the metal matrix composites (MMCs) [2]. Light metal matrix composites are valued, particularly in certain applications where low density should be balanced with mechanical requirements [3]. In particular, titanium composites (TMCs) offer these advantages over other light metal matrices [1, 4]. Their good corrosion behaviour and high specific properties make TMCs one of the most suitable candidates for aerospace applications [5].

Several authors have described several classifications of these materials. One of these classifications could be considered based on the kind of reinforcements: continuous or discontinuous reinforcement [6, 7]. Other classifications could be done according to the manufacturing route: traditional methods or powder metallurgy techniques [8–12]. An interesting route to promote the strengthening of the matrix is the "in situ" formation of secondary phases. This method allows a clean and well bounding between the matrix and the reinforcement [9]; consequently, better final behaviour of the TMCs may be expected [13, 14].

On the basis of previous and recent studies, this work focuses on TMCs which were manufactured employing discontinuous ceramic reinforcement. These ceramic phases were selected according to their reactivity with the titanium matrix. Many authors show the great variety of ceramic reinforcements; however, in this investigation TiC, TiB_2, B, and B_4C particles were studied. They were considered as precursors of secondary phase formation by in situ techniques [15–20].

From the manufacturing process point of view, powder metallurgy techniques of hot consolidation have proved useful at the study of in situ composites [21]. Therefore, for the development of TMCs, inductive hot pressing has been selected among other manufacturing processes. The experience of the authors in this technique had led to laying the basis of this study [22–28].

By a thorough analysis of the properties of the produced specimens via inductive hot pressing at different temperatures by the use of several starting material compositions, specific features of the TMCs could be studied. In this regard, the employment of the XRD technique is crucial in understanding the reaction phenomena between the matrix and the reinforcement. Furthermore, the behaviour of the ceramic particles in the matrix could be unpredictable and variable depending upon many factors; this study is the main objective to analyse the phenomena that could occur when factors as starting powder composition and processing parameters are varied and ultimately how these factors affect the final properties of the TMCs.

2. Materials and methods

The interest in understanding the influence of the starting materials on the final behaviour of TMCs motived the study of three ceramic materials as reinforcements testing various concentrations in titanium matrices. Hence, for the TMC manufacture, diverse starting blends were prepared. The innovation of this investigation lies in the phase analysis carried out in specimens made from these blends. The employment of several operation temperatures and reinforcement typologies and concentrations allowed for the searching of significant differences, among the fabricated TMCs, while all these specimens have been processed at similar processing conditions.

The titanium grade 1 employed was manufactured by TLS GmbH (Bitterfeld, Germany). This titanium powder showed a spherical morphology and D_{50} below 45 μm. The four ceramic reinforcements were chosen considering their reactive behaviour in respect of the secondary phases' formation in titanium matrices. The supplier for TiC powder was H.C. STARK GmbH (Goslar, Germany) and for B_4C powders was abcr GmbH (Karlsruhe, Germany), and the company for TiB_2 was Treibacher Industrie AG (Althofen, Austria). The characterization of all the powders was carried out to verify the manufacturers' data about their size and morphology. The particle size distribution of the powders was measured by laser diffraction analysis (Mastersizer 2000, Malvern Instruments, Malvern, United Kingdom); these results are shown in **Table 1**.

X-ray powder diffraction analysis was done using a Bruker D8 Advance A25 (Billerica, Massachusetts, United States of America) with Cu-K_α radiation for the phase characterisation of as-received Ti, TiC, TiB_2, and B_4C powders and studying

Material	Morphology	D_{50} (μm)
Ti	Spherical	29.05
TiC	Faceted	4.9
TiB_2	Irregular	4.76
B_4C	Faceted	63.76

Table 1.
Characteristics of the starting materials.

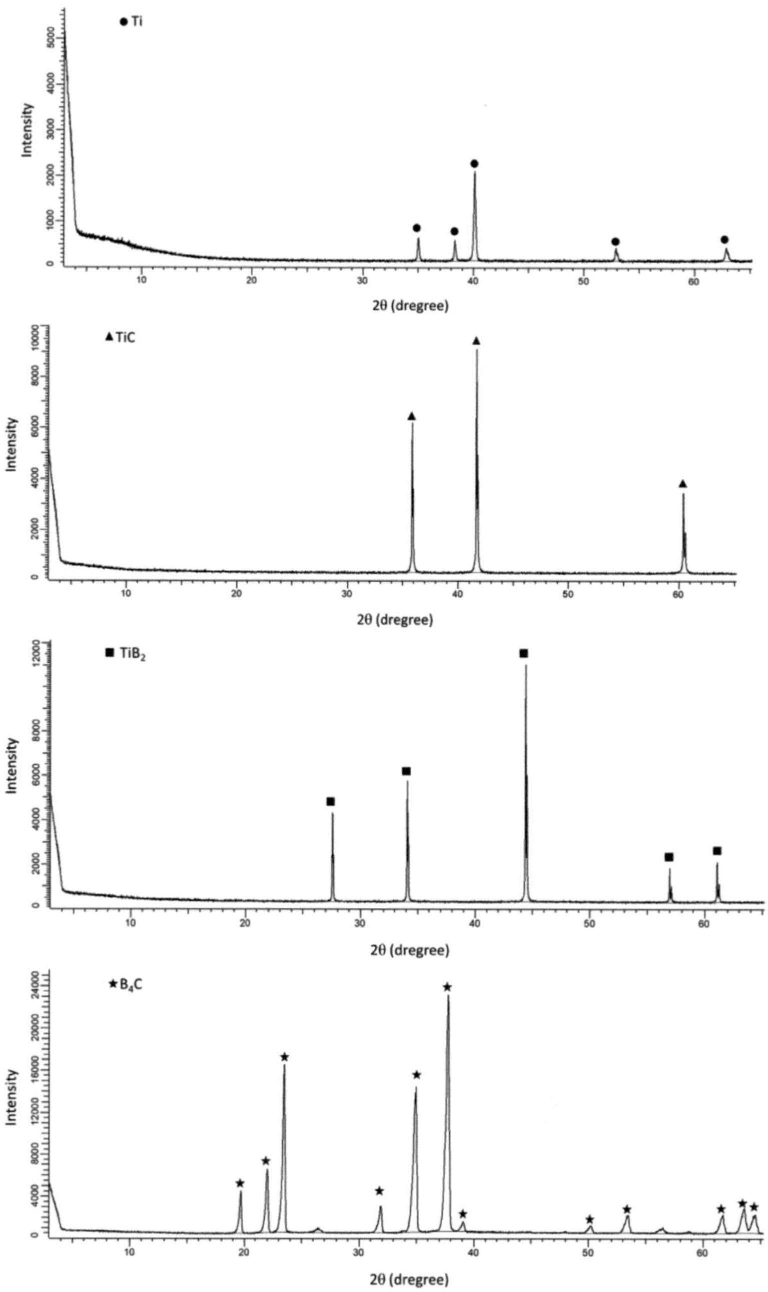

Figure 1.
XRD analysis of the starting powders: Ti grade 1, TiC, TiB_2, and B_4C.

the phase evolution of sintered TMCs. The reference intensity ratio (RIR) analysis was performed to semi-quantitatively determine the phases. This method is based on setting the diffraction data to the diffraction of standard reference materials. The intensity of a diffraction peak profile is a convolution of diverse factors, being the most representative of the concentration of the analysed species.

In **Figure 1**, the X-ray diffraction spectra of the starting materials (as-received) are shown. Based on the obtained diffraction patterns, these materials consist of only Ti, TiC, TiB_2, and B_4C, respectively.

Previously in the TMC consolidation, the raw material blends were prepared according to the fixed percentages in volume (see **Table 2**). The mixing procedure was described in previous authors' work [28]. Next, the specimens were sintered. To consolidate the TMCs, a self-made hot pressing machine, inductive hot pressing (iHP) equipment of RHP-Technology GmbH & Co. KG (Seibersdorf, Austria), was used. This machine enabled the operational cycle time to be reduced thanks to its advantageous high heating rate, which in turn is due to its special inductive heating setup. The prepared powders were inserted in the graphite die; it was lined with thin graphite paper and a protective coating of boron nitride (BN). The same procedure and die were used for all the iHP cycles (punch Ø 20 mm). Methods based on this rapid hot consolidation are considered as preferred techniques for in situ fabrication of nearly fully dense TMCs [29]. In **Table 2**, the processing conditions are shown. Likewise, the curves of the process cycle are represented graphically in

Ti matrix and reinforcement	**Processing parameters**			
Reinforcement material	Volume [%]	Temperature [°C]	Time [min]	Pressure [MPa]
TiC	10, 20, 30	1000, 1100, 1200	15	50
TiB_2	10, 20, 30	1000, 1100, 1200	15	50
B_4C	10, 20, 30	1000, 1100, 1200	15	50

Table 2.
Reinforcement percentages and processing parameters.

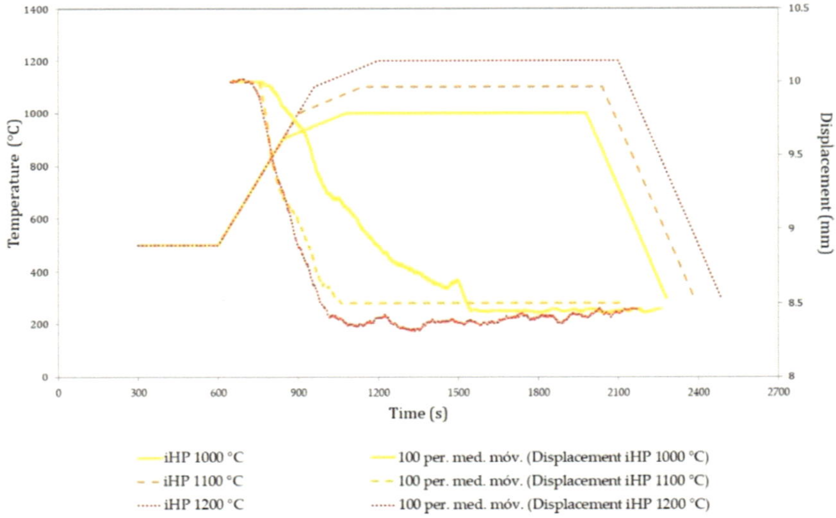

Figure 2.
Diagram of the inductive hot-pressing cycle: time vs. temperature and piston displacement.

Figure 2. The iHP equipment worked in vacuum conditions (5 10^{-4} bar), the cycle heating rate being 50°C/min. Following the consolidation, the specimens were dislodged from the graphite die and cut in half vertically.

After a thorough metallographic preparation on the cross section of the specimens, the X-ray analysis and the microstructural study were performed. It was studied by means of SEM, using JEOL 6460LV (Tokyo, Japan) and FEI Teneo (Oregon, United States of America). Furthermore, the hardness measurement was carried out; seven indentations were performed on each specimen, avoiding the ceramic particles. A tester model, Struers Duramin A300 (Ballerup, Denmark), was employed to ascertain the Vickers hardness (HV2). An ultrasonic method (Olympus 38DL, Tokyo, Japan) was employed to calculate the Young modulus by measuring longitudinal and transversal propagation velocities of acoustic waves [30]. Archimedes' method (ASTM C373-14) was set for the density measurement.

3. Results and discussions

3.1 X-ray diffraction analysis and microstructural study

This section has been structured according to the employed reinforcement, in order to present the results and to perform their discussion clearly and concisely. Therefore, there have been three subparts taking into account the ceramic materials used as starting reinforcements in the TMC manufacturing.

3.1.1 TiC

The X-ray diffraction spectra of TMCs made from TiC-Ti blends are shown in **Figure 3**. Based on the obtained diffraction patterns, these materials consist of Ti and TiC_x phases. On the one hand, the X-ray analysis reveals that there are only Ti and TiC phases in specimens produced at 1000°C, regardless of whether the highest or the lowest TiC concentration (vol.%) was used in the starting blend. Likewise, only TiC stoichiometric phase is detected in specimens made from 10 vol.% of TiC, even in specimens produced at 1100 and 1200°C. On the other hand, the peak intensity of the Ti phase decreased; meanwhile, there was an apparition of slight diffraction peaks of nonstoichiometric TiC phase named $TiC_{0.67}$. It suggested that there were possible reactions between the Ti and diffused C from the TiC particles at high concentrations (20 vol.% TiC and 30 vol.% TiC). The intensification of the nonstoichiometric TiC peaks from 1100 to 1200°C indicates the increase in the volume fraction of this phase, which can be confirmed by the RIR semi-quantification analysis. The results from RIR analysis are shown in **Table 3**. These results may lead to the assumption that the phenomenon of C diffusion was more significant at the highest TiC concentration (30 vol.%) and the highest operational temperature (1200°C). In agreement with the values of the RIR semi-quantification analysis in **Table 3**, the higher the TiC in starting materials was used, the higher the TiC phase values in the RIR analysis was detected.

To clarify the distribution of the nonstoichiometric TiC_x phases in TMCs, energy-dispersive X-ray spectroscopy (EDS) analysis was performed. In **Figure 4**, the EDS result revealed that there were concentration gradients between the centres of the TiC particles and the matrix. This clearly demonstrated the value of the temperature and the starting material compositions as influencing factors in the final behaviour of the TMCs. Moreover, it can be observed from **Figure 4** that C was diffused in the region, which is rich in titanium. This is in accordance with the slight displacement of the Ti peaks in the TMC patterns when the specimens were consolidated especially at 1200°C.

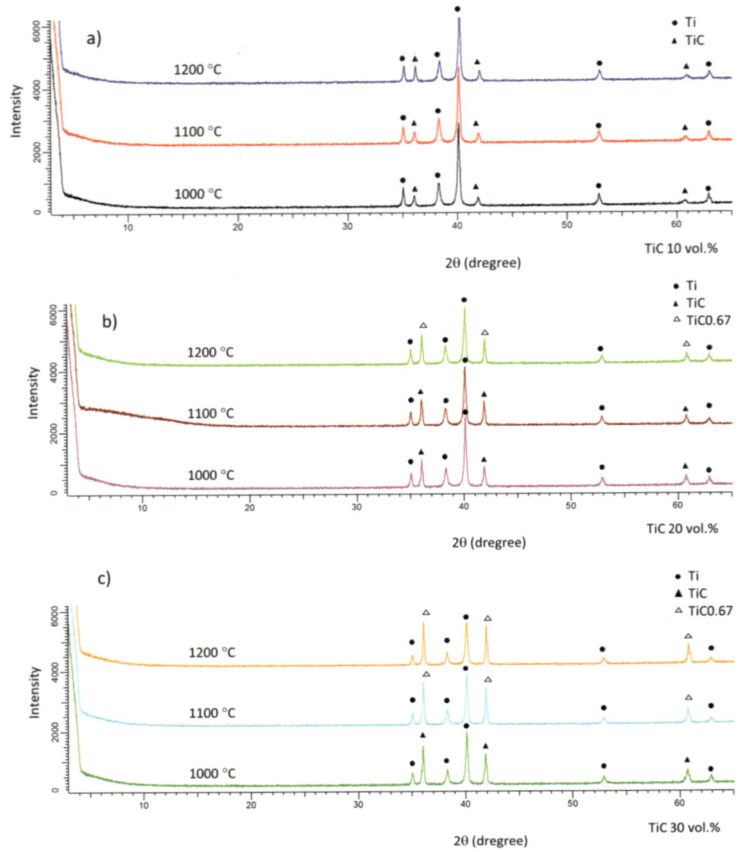

Figure 3.
XRD patterns of TMCs reinforced using (a) 10 vol.% of TiC, (b) 20 vol.% of TiC and (c) 30 vol.% of TiC.

Temperature (°C)	vol.%	Ti (%)	TiC (%)	$TiC_{0.67}$ (%)
1000	10	97.8	2.2	
	20	96.3	6.4	
	30	88.5	11.5	
1100	10	96.9	3.1	
	20	91.9	8.1	
	30	84.4		15.6
1200	10	94.6	3.6	1.8
	20	91.4		8.6
	30	81.0		19.0

Table 3.
Reinforcement percentages and processing parameters.

From the microstructural point of view, there were several differences observed in the specimens, which depended not only on the processing temperature employed but also on the starting reinforcement concentration. In this regard, the lower the concentration of TiC (10 vol.%), the fewer the pores observed in the TMC microstructure. Moreover, some agglomerations of the TiC particles could be clearly appreciated in specimens made from the blend with 30 vol.% of TiC; there are little pores observed

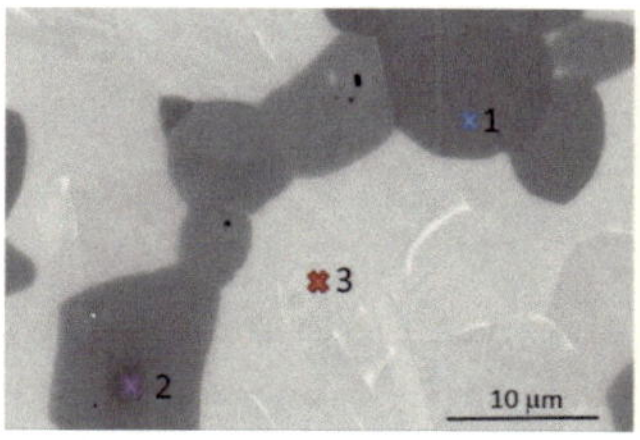

	Elements [Atomic %]	
Spots	Ti	C
1	60.94	39.06
2	51.36	48.64
3	85.51	14.49

Figure 4.
On the left, SEM image of TMCs processed at 1200°C, with starting TiC composition of 20 vol.%. On the right, EDX analysis of three spots on the cross section of such TMC.

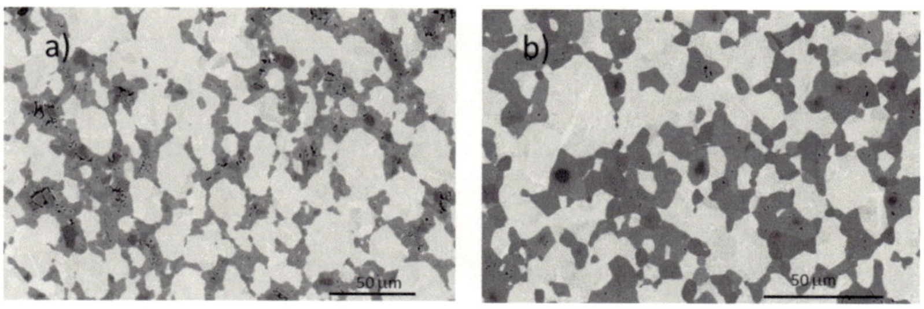

Figure 5.
SEM images of TMCs made from composition of 30 vol.% of TiC hot consolidated at (a) 1000°C and (b) 1200°C.

in the centre of these ceramic particles' agglomerations in the titanium matrices (**Figure 5a**). The pores tended to close with the increase in temperature. In line with the diffusion phenomenon mentioned above, a possible reason for the porous reduction may be the diffusion of the C element and, consequently, the formation of TiC_x phases. Furthermore, the cited pores could also be caused by material removal during the metallographic preparation, which suggested that major bonding between TiC particles and the matrix decreases the material removal. It indicates that the reaction between the C sourced by TiC particles and Ti from the matrix involved a strong interfacial bonding. Therefore, the rising of the temperature benefited, and it was very useful to obtain major densification. It is worth noting that the reinforcement agglomeration could be a problem as a barrier for affecting the diffusion phenomenon and the interfacial contact. For that reason, the pores are only observed in the centre of the mentioned agglomerations (see **Figure 5**).

3.1.2 TiB_2

Figure 6 shows the XRD patterns of the TMCs reinforced with TiB_2 particles. In this respect, particular attention will be devoted to the existence of peaks of Ti_3B_4, while there was an increment of the temperature from 1100 to 1200°C. Likewise, it can be seen that the XRD patterns of the specimens produced at 1000°C only contain strong diffraction peaks of TiB_2 phase and slight diffraction peaks of TiB phase. The Ti_3B_4 peaks appear independently of the starting TiB_2 concentration (vol.%), being only related to the processing temperature (1100 and 1200°C).

Table 4 shows the semi-quantification analysis of the TMCs reinforced by TiB_2 particles. As many authors describe [9, 12, 13, 16, 19, 22, 29], there are reactions between B from TiB_2 particles and the Ti matrix, resulting in the in situ TiB phase. Thus, it would be expected that the percentages of in situ TiB phase were proportional to the initial composition of TiB_2 in the starting blend. However, observing the values presented in **Table 4**, the key parameter was the temperature instead of the concentration,

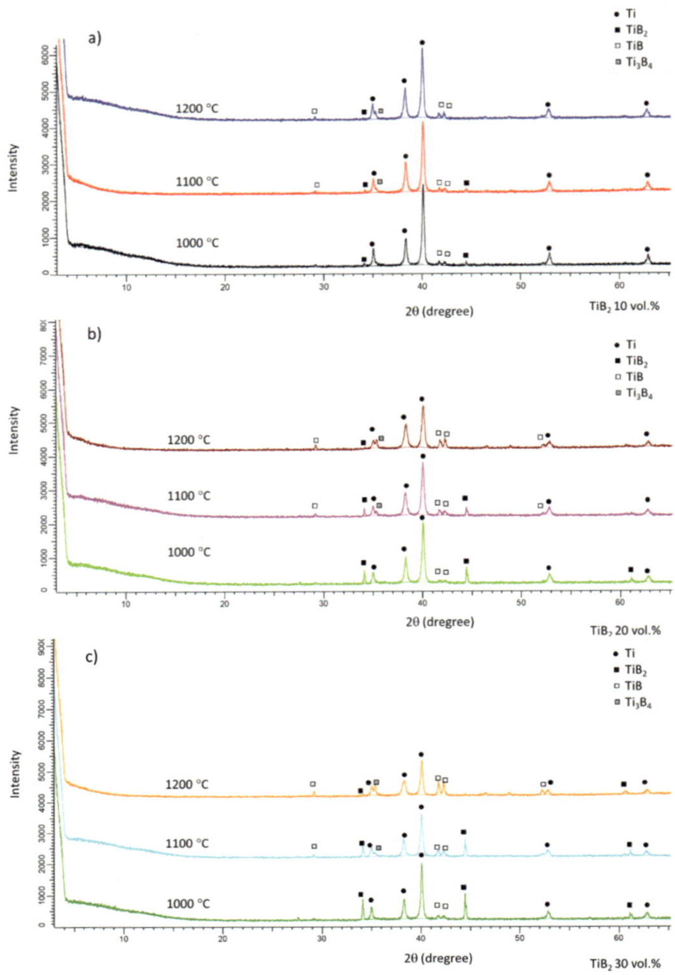

Figure 6.
XRD patterns of TMCs reinforced in the starting blend with (a) 10 vol.% of TiB_2, (b) 20 vol.% of TiB_2, and (c) 30 vol.% of TiB_2.

Ti matrix and TiB_2					
Temperature [°C]	vol.%	Ti (%)	TiB_2 (%)	TiB	Ti_3B_4 (%)
1000	10	91.3	4.7	4.0	
	20	78.2	19.0	2.8	
	30	68.3	28.1	2.9	
1100	10	90.2	4.1	3.9	1.8
	20	77.9	12.0	6.8	3.3
	30	65.0	24.8	7.4	2.8
1200	10	88.5	2.5	6.2	2.6
	20	76.7	3.9	12.3	7.1
	30	72.9	3.5	20.9	3.4

Table 4.
RIR semi-quantification analysis of TMCs made from Ti-TiB_2 blends, manufactured at different temperatures (by iHP).

promoting the apparition of TiB as in situ formed phase. Owing to the rising temperature, the diffusion mechanism was driven by the temperature increments of 100°C (from 1000 to 1100°C and from 1100 to 1200°C). The highest temperature (1200°C) played a major role in the formation of TiB, independently of the operational temperature. Obviously, at the same temperature, there was more TiB detected in specimens made from starting powder with the higher TiB_2 composition (30 vol.% of TiB_2).

Microstructural study of these TMCs confirmed the visual existence of the in situ TiB_x phases. Moreover, some pores were detected in areas where the TiB_2 particles were slightly agglomerated. As mentioned in the results of the microstructural analysis of TMC reinforced by TiC_x phases, the referred pores were located in the centre of particle agglomeration. The higher the concentration of particles and the lower the operational temperature, the more significant the apparition of pores in the TMCs. In **Figure 7**, the commented pores can be recognized.

The influence of the temperature was relevant once again to close these pores, as in similitude with the TiC. Many studies [31] attempted to show the importance of strong bonding between the matrix and the TiB_x phases; the no contact between the reinforcement and the matrix, in addition to the inappropriate processing temperature, inhibited the formation of in situ secondary phases. By increasing the operational temperature, improvement in the diffusion phenomenon was expected.

SEM images of the microstructure of TMCs processed at 1100°C are shown in **Figure 8**. The results about homogenous distribution and increase in the volume of reinforcements in the Ti matrix are in accordance to the RIR analysis. In **Figure 8a**, the reinforcements on the matrix can be easily recognized. Observing the microstructural evolution by increment of the composition, the smaller TiB_2 particles were surrounded by the in situ formed phases when the starting composition of TiB_2 was the lowest. However, in **Figure 8c**, coarse TiB_2 particles were also surrounded by phases with minor size.

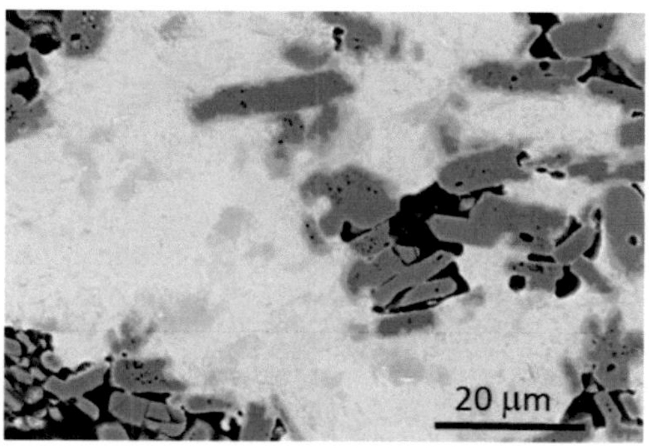

Figure 7.
SEM image of TMC reinforced with 30 vol.% of TiB_2 particles consolidated at 1000°C.

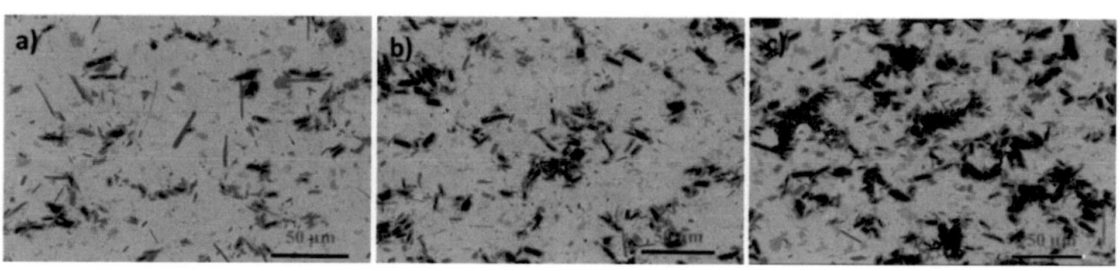

Figure 8.
SEM images of TMCs processed at 1100°C with different percentages of TiB_2 in the starting blends: (a) 10 vol.%, (b) 20 vol.%, and (c) 30 vol.%.

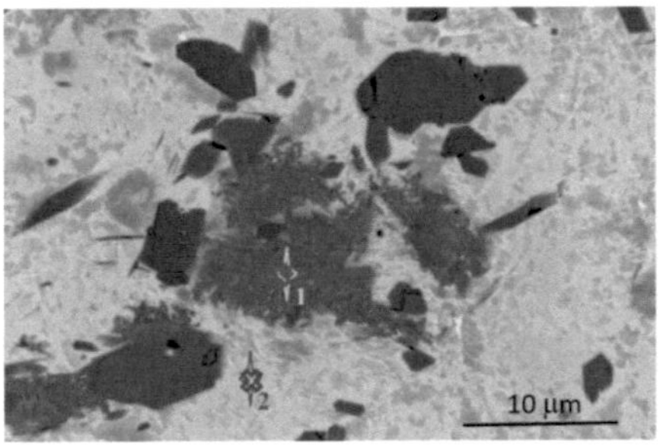

	Elements [Atomic %]	
Spots	Ti	B
1	33.93	66.07
2	55.05	44.95

Figure 9.
SEM image of TMC with 10 vol.% of TiB_2 in the starting blend, processed at 1200°C.

The rising in temperature was crucial for reactions between the matrix and the TiB_2 particles. At 1200°C, there were major diffusion of B through the matrix and more formation of the in situ TiB_x phases. **Figure 9** shows two different areas on a cross section (iHP at 1200°C), where the B distribution varied considerably; the darkest region in the centre corresponds with the highest concentration of B. It suggests that the dark grey areas were originally the TiB_2 particles, which were surrounded by the in situ TiB_x phases.

3.1.3 B_4C

The use of B_4C offers considerable scope for diversification and development of in situ secondary phases (TiC and TiB). Hence, a wide range of studies intended to demonstrate the suitability of B_4C as a source of B and C for in situ secondary phases, owing to its reactive behaviour with the Ti matrix. The B_4C particles can trigger reactions whose products contribute to enhance the TMC properties. In this regard, TiC and TiB phases may expect to be observed and analysed in this type of TMCs. **Figure 9** shows the XRD patterns of the TMCs reinforced with B_4C particles. It can be verified that the highest temperature of the iHP process and the holding time (15 minutes) were insufficient for a full reaction between the boron carbide particles and the titanium matrices, even at the lowest concentration of B_4C. Thus, this fact occurred independently of the starting compositions, confirmed by the

Ti matrix and B_4C					
Temperature [°C]	B_4C vol.%	Ti (%)	B_4C (%)	TiB (%)	TiC (%)
1000	10	92.2	5.7	1.6	0.5
	20	86.4	10.8	1.8	1.0
	30	78.0	18.3	2.2	1.5
1100	10	90.6	5.7	2.7	1.0
	20	81.2	10.3	6.5	2.0
	30	73.3	17.9	6.5	2.3
1200	10	89.0	4.9	5.0	1.1
	20	80.6	10.0	7.0	2.3
	30	64.4	17.1	14.3	4.2

Table 5.
RIR semi-quantification analysis of TMCs made from Ti-B_4C blends, manufactured at different temperatures (by iHP).

existence of peaks related to the boron carbide. Likewise, there were observed peaks matching TiB and TiC patterns.

The intensification of TiB and TiC peaks from 1000 to 1200°C reflects the increase in the volume fraction of these phases, which can also be seen in the RIR analysis shown in **Table 5**.

The microstructural study shows the homogenous dispersion of the B_4C particles in the matrix. In this context, there were no agglomerations visually detected. This suggests that there was no porosity related to particles agglomerations as commented previously in TMCs reinforced by TiB_2 and TiC. It could be considered as an advantage of the B_4C as reinforcement in comparison with other ceramic particles. **Figure 11** shows TMCs processed at the same temperature with different B_4C percentages.

Regarding the processing temperature, there were significant differences related to the reaction between the matrix and the B and C from the B_4C particles. At the lowest temperature (1000°C), the formation of the in situ TiB and TiC phases was proportional to the starting content of B_4C. Employing 10 vol.% of B_4C, small proportions of in situ phases were detected (see **Table 5**). However, increasing the temperature to 1100°C and using 10 vol.% of B_4C, the percentage of in situ TiC phase doubled its value, also, by the employment of 20 and 30 vol.%. This is in agreement with the intensity of the peaks of this phase in the TMC patterns (**Figure 10**). As expected, the major in situ formation of secondary phases took place at 1200°C.

Figure 12 reveals how the in situ phases surrounded the B_4C particles, being a reaction layer clearly defined. Obviously, the higher the starting B_4C composition, the more

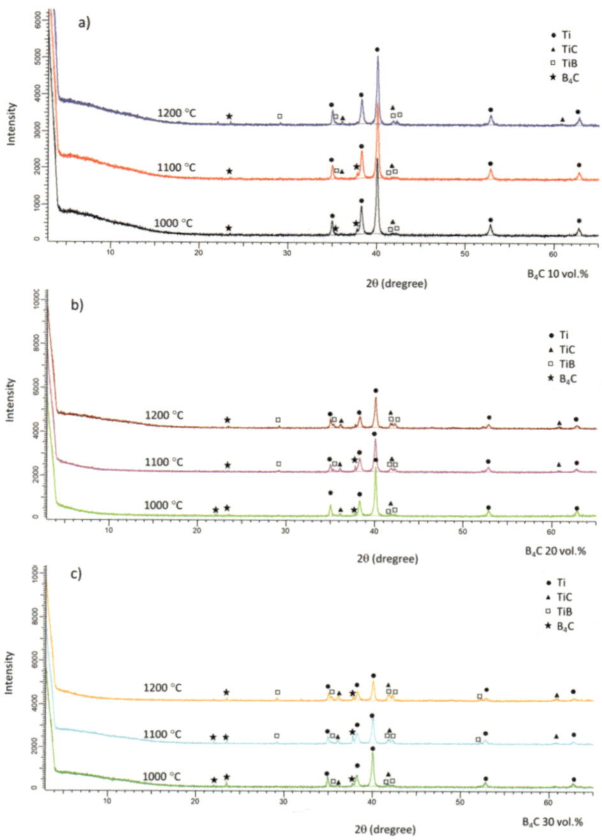

Figure 10.
XRD patterns of TMCs reinforced in the starting blend with (a) 10 vol.% of B_4C, (b) 20 vol.% of B_4C, and (c) 30 vol.% of B_4C.

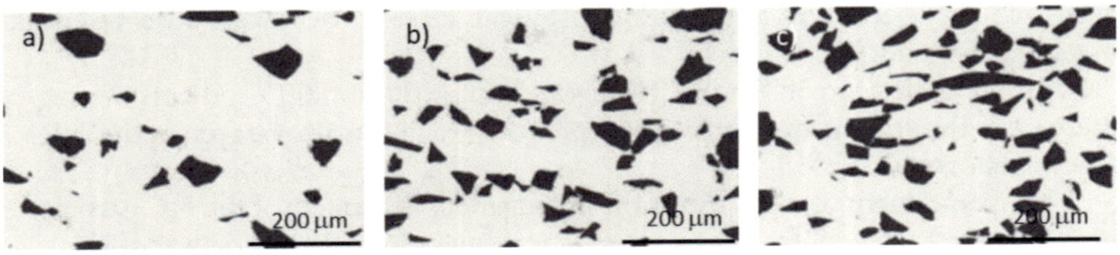

Figure 11.
SEM images of TMCs produced at 1000°C with (a) 10 vol.% of B_4C, (b) 20 vol.% of B_4C, and (c) 30 vol.% of B_4C.

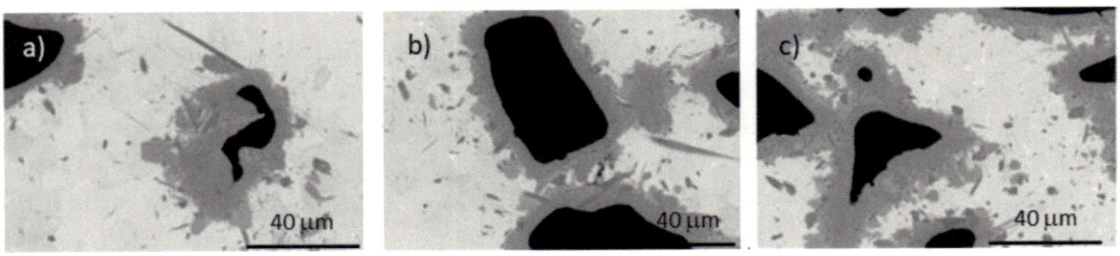

Figure 12.
SEM images of TMCs produced at 1200°C with (a) 10 vol.% B_4C, (b) 20 vol.% B_4C, and (c) 30 vol.% B_4C.

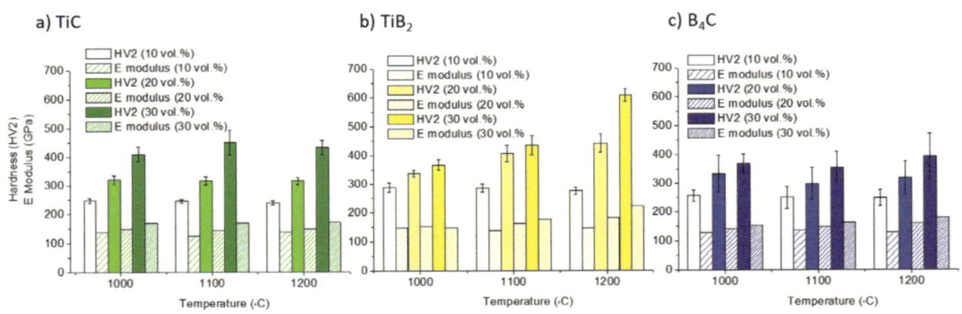

Figure 13.
Hardness (HV2) and Young modulus values vs. operational temperature of the TMCs made from the different blends.

the formation of in situ phases. Regardless of the starting compositions, the morphologies of the in situ phases TiC and TiB are similar to the ones observed previously. On the one hand, there were precipitates with the particular whisker shape of TiB in the matrix. On the other hand, the presence of TiC can be seen as globular precipitated; both in situ phase morphologies have been wide and thoroughly studied [32].

3.2 Physical properties of the TMCs.

The relative density of the specimens was around 99.5% in the majority of the specimens, even in those whose microstructures had a few pores. It means that the processing parameters were suitable to achieve full densification.

As expected, the highest values of hardness and Young modulus were recorded in specimens whose starting blends were made from the highest ceramic particle contents. **Figure 13** shows a comparison of the hardness and Young modulus values of the TMCs produced at the three processing temperatures (1000, 1100, and 1200°C) and using the three compositions (10, 20, and 30 vol.%).

The operational temperature contributed to enhancing the hardness and the Young modulus; however, its influence varied depending on the type of ceramic particles employed in the starting blend, as reflected in **Figure 13**. TMCs reinforced by

TiB-TiB_2 phases showed the highest hardness measured. This development is closely related to the content of in situ formed TiB. Although there was also in situ formed TiB in TMCs made from blends with B_4C particles, the maximum percentage formed (14 vol.%) in these specimens was lower than in TMCs made from the blend with TiB_2 (20 vol.%). In both cases, the TMCs were processed at 1200°C and 30 vol.%. In similar conditions, the highest Young modulus was also observed in TMCs reinforced with TiB-TiB_2 phases, in agreement with the commented results above.

In specimens made from blends with TiC, the main variation was only caused by the addition of more TiC content. Hardness and Young modulus values hardly increased by temperature, despite the diffusion of the C in the matrix and the $TiC_{0.67}$ formed.

Contrary to common thinking, the B_4C reinforcement did not behave as the best precursor of in situ phases. Consequently, the expected properties may vary from the obtained values of hardness and Young modulus. The TiC and TiB formed were slightly lower than the in situ TiB formed in TMCs with TiB_2. That means that the diffusion of B alone was major and the C could decelerate such diffusion. Furthermore, it should be highlighted that in specimens made from B_4C, the values of hardness and Young modulus showed a wide standard deviation. This could be related to the in situ formed precipitates and their dispersion in the matrix.

4. Conclusions

The conclusions of the current study which analyse the influence of the starting materials and operational temperature in the TMC properties are drawn:

- Reinforcing the titanium matrix with ceramic materials results in an enhancement of the TMC mechanical properties caused by the formation of in situ phases.

- XRD analysis states that the diffusion phenomenon of B and C elements into the matrix increases by the rising temperature; it is becoming increasingly important in the apparition of secondary phases.

- In evaluating the appropriateness of the operational parameters, the lower the temperature, the less the reactivity reinforcement matrix. This phenomenon was more significant when the concentration of reinforcement was the lowest one.

- The highest hardness and Young modulus of the TMCs were measured in specimens reinforced by TiB_2 particles.

- The densification of the specimens was achieved at the processing parameters tested.

Acknowledgements

The authors gratefully acknowledge the company "RHP-Technology GmbH" and the managers Dr. Neubauer and Dipl. Eng. Kitzmantel for their partial financial support of this work. In addition, the authors want to thank the Universidad de Sevilla for the use of experimental facilitates at CITIUS Microscopy and X-Ray Laboratory Services (VI PPIT-2018-I.5 EVA MARÍA PÉREZ SORIANO).

Conflict of interest

The authors declare no conflict of interest.

Author details

Eva María Pérez-Soriano, Cristina M. Arévalo-Mora
and Isabel Montealegre-Meléndez*
Universidad de Sevilla, Sevilla, Spain

*Address all correspondence to: imontealegre@us.es

References

[1] Chawla KK. Composite Materials: Science and Engineering. 2nd ed. New York: Springer; 2010. DOI: 10.1007/978-0-387-74365-3. 542 p

[2] Tjong S, Ma Z. Microstructural and mechanical characteristics of in situ metal matrix composites. Materials Science & Engineering R: Reports. 2000;**29**(3):49-113. DOI: 10.1016/S0927-796X(00)00024-3

[3] Lawrance GA. Front matter. In: Lawrance GA, editor. Introduction to Coordination Chemistry. Chichester: Wiley; 2010. pp. i-xix. DOI: 10.1002/9780470687123

[4] Smith PR, Froes FH. Developments in titanium metal matrix composites. Journal of Metals. 1984;**36**(3):19-26. DOI: 10.1007/BF03338403

[5] Leyens C, Peters M, editors. Titanium and Titanium Alloys: Fundamentals and Applications. 1st ed. Köln: Wiley; 2003. DOI: 10.1002/3527602119. 532 p

[6] Kaczmar JW, Pietrzak K, Włosiński W. The production and application of metal matrix composite materials. Journal of Materials Processing Technology. 2000;**106**(1):58-67. DOI: 10.1016/S0924-0136(00)00639-7

[7] Ranganath SA. Review on particulate-reinforced titanium matrix composites. Journal of Materials Science. 1997;**32**(1):1-16

[8] Gofrey TMT, Goodwin PS, Ward-Close CM. Titanium particulate metal matrix composites: Reinforcement, production methods, and mechanical properties. Advanced Engineering Materials. 2000;**2**(3):85-91. DOI: 10.1002/(SICI)1527-2648(200003)2:3%3C85:AID-ADEM85%3E3.0.CO;2-U

[9] Radhakrishna Bhat BV, Subramanyam J, Bhanu Prasad VV. Preparation of Ti-TiB-TiC & Ti-TiB composites by in-situ reaction hot pressing. Materials Science and Engineering A. 2002;**325**(1):126-130. DOI: 10.1016/S0921-5093(01)01412-5

[10] Campbell FC. Metal matrix composites. In: Campbell FC, editor. Manufacturing Technology for Aerospace Structural Materials. Amsterdam: Elsevier Science; 2006. pp. 419-457. DOI: 10.1016/B978-1-85617-495-4.X5000-8

[11] Montealegre Meléndez I, Neubauer E, Danninger H. Consolidation of titanium matrix composites to maximum density by different hot pressing techniques. Materials Science and Engineering A. 2010;**527**(16-17): 4466-4473. DOI: 10.1016/j.msea.2010.03.093

[12] Zhang C, Kong F, Xiao S, Niu H, Xu L, Chen Y. Evolution of microstructural characteristic and tensile properties during preparation of TiB/Ti composite sheet. Materials and Design. 2012;**36**:505-510. DOI: 10.1016/j.matdes.2011.11.060

[13] Tjong SC, Mai YM. Processing-structure-property aspects of particulate- and whisker-reinforced titanium matrix composites. Composites Science and Technology. 2008;**68**(3-4):583-601. DOI: 10.1016/j.compscitech.2007.07.016

[14] Zhang Z, Qin J, Zhang Z, Chen Y, Lu W, Zhang D. Microstructure effect on mechanical properties of in situ synthesized titanium matrix composites reinforced with TiB and La_2O_3. Materials Letters. 2010;**64**(3):361-363. DOI: 10.1016/j.matlet.2009.11.019

[15] Xu D, Lu WJ, Yang ZF, Qin JN, Zhang D. In situ technique for synthesizing multiple ceramic particulates reinforced titanium

matrix composites (TiB+TiC+Y_2O_3)/Ti. Journal of Alloys and Compounds. 2005;**400**(1):216-221. DOI: 10.1016/j.jallcom.2005.04.018

[16] Ni DR, Geng L, Zhang J, Zheng ZZ. Effect of B_4C particle size on microstructure of in situ titanium matrix composites prepared by reactive processing of Ti–B_4C system. Scripta Materialia. 2006;**55**(5):429-432. DOI: 10.1016/j.scriptamat.2006.05.024

[17] Ma F, Lu W, Qin J, Zhang D, Ji B. Effect of forging and heat treatment on the microstructure of in situ TiC/Ti-1100 composites. Journal of Alloys and Compounds. 2007;**428**(1):332-337. DOI: 10.1016/j.jallcom.2006.03.060

[18] Zhang Y et al. Damping capacity of in situ TiB_2 particulates reinforced aluminium composites with Ti addition. Materials and Design. 2007;**28**(2):628-632. DOI: 10.1016/j.matdes.2005.07.015

[19] Jimoh A. In-situ particulate-reinforcement of titanium matrix composites with borides [thesis]. Johannesburg: University of the Witwatersrand; 2010. Available from: http://hdl.handle.net/10539/9323

[20] Zhang Y, Sun J, Vilar R. Characterization of (TiB+TiC)/TC4 in situ titanium matrix composites prepared by laser direct deposition. Journal of Materials Processing Technology. 2011;**211**(4):597-601. DOI: 10.1016/j.jmatprotec.2010.11.009

[21] Jimoh A, Sigalas I, Hermann M. In situ synthesis of titanium matrix composite (Ti-TiB-TiC) through sintering of TiH_2-B_4C. Materials Sciences and Applications. 2012;**03**(01):30-35. DOI: 10.4236/msa.2012.31005

[22] Montealegre-Meléndez I, Neubauer E, Arévalo C, Rovira A, Kitzmantel M. Study of titanium metal matrix composites reinforced by boron carbides and amorphous boron particles produced via direct hot pressing. Key Engineering Materials. 2016;**704**:85-93. DOI: 10.4028/www.scientific.net/KEM.704.85

[23] Arévalo C, Kitzmantel M, Neubauer E, Montealegre-Meléndez I. Development of Ti-MMCs by the use of different reinforcements via conventional hot-pressing. Key Engineering Materials. 2016;**704**:400-405. DOI: 10.4028/www.scientific.net/KEM.704.400

[24] Arévalo C, Montealegre-Meléndez I, Ariza E, Kitzmantel M, Rubio-Escudero C, Neubauer E. Influence of sintering temperature on the microstructure and mechanical properties of In situ reinforced titanium composites by inductive hot pressing. Materials. 2016;**9**(11):919. DOI: 10.3390/ma9110919

[25] Neubauer E, Vály L, Kitzmantel M, Grech D, Ortega AR, Montealegre-Meléndez I, et al. Titanium matrix composites with high specific stiffness. Key Engineering Materials. 2016;**704**:38-43. DOI: 10.4028/www.scientific.net/KEM.704.38

[26] Montealegre-Meléndez I, Arévalo C, Perez-Soriano E, Neubauer E, Rubio-Escudero C, Kitzmantel M. Analysis of the influence of starting materials and processing conditions on the properties of W/Cu alloys. Materials. 2017;**10**(2):142. DOI: 10.3390/ma10020142

[27] Ariza E, Montealegre-Meléndez I, Arévalo C, Kitzmantel M, Neubauer E. Ti/B_4C composites prepared by in situ reaction using inductive hot pressing. Key Engineering Materials. 2017;**742**:121-128. DOI: 10.4028/www.scientific.net/KEM.742.121

[28] Montealegre-Meléndez I, Arévalo C, Pérez-Soriano EM, Kitzmantel M, Neubauer E. Microstructural and XRD analysis and study of the properties of the system Ti-TiAl-B_4C processed

under different operational conditions. Metals. 2018;**8**(5):367. DOI: 10.3390/ma10111240

[29] Sabahi Namini A, Dilawary SAA, Motallebzadeh A, Shahedi Asl M. Effect of TiB_2 addition on the elevated temperature tribological behavior of spark plasma sintered Ti matrix composite. Composites. Part B, Engineering. 2019;**172**:271-280. DOI: 10.1016/j.compositesb.2019.05.073

[30] Ahmad A, Bond LJ, editors. ASM Handbook. Nondestructive evaluation of materials. 9th ed. Vol. 17. Ohio, United State of America; 1989

[31] Wang T, Gwalani B, Shukla S, Frank M, Mishra RS. Development of in situ composites via reactive friction stir processing of Ti–B_4C system. Composites. Part B, Engineering. 2019;**172**:54-60. DOI: 10.1016/j.compositesb.2019.05.067

[32] Rielli VV, Amigó-Borrás V, Contieri RJ. Microstructural evolution and mechanical properties of in-situ as-cast beta titanium matrix composites. Journal of Alloys and Compounds. 2019;**778**:186-196. DOI: 10.1016/j.jallcom.2018.11.093